NORTH KOREA'S SECOND NUCLEAR CRISIS AND NORTHEAST ASIAN SECURITY

North Korea's Second Nuclear Crisis and Northeast Asian Security

Edited by
SEUNG-HO JOO
University of Minnesota-Morris, USA

TAE-HWAN KWAK
Eastern Kentucky University, USA

ASHGATE

Published by
Ashgate Publishing Limited
Gower House
Croft Road
Aldershot
Hampshire GU11 3HR
England

Ashgate Publishing Company
Suite 420
101 Cherry Street
Burlington, VT 05401-4405
USA

Ashgate website: http://www.ashgate.com

British Library Cataloguing in Publication Data
North Korea's second nuclear crisis and Northeast Asian
security
1. Nuclear weapons - Korea (North) 2. Nuclear arms control
- Korea (North) 3. Korea (North) - Foreign relations
4. Korea (North) - Foreign relations - United States
5. United States - Foreign relations - Korea (North)
6. United States - Foreign relations - 2001-
327.5'193'0090511

Library of Congress Cataloging-in-Publication Data
North Korea's second nuclear crisis and northeast Asian security / edited by Seung-Ho Joo and Tae-Hwan Kwak.
p. cm.
Includes bibliographical references and index.
ISBN 978-0-7546-7176-3
1. Nuclear arms control--Korea (North) 2. Nuclear weapons--Korea (North) 3. Six-party Talk. 4. Korea (North)--Foreign relations. 5. Korea (North)--Foreign relations--Korea (South) 6. Korea (South)--Foreign relations--Korea (North) I. Joo, Seung-Ho, 1959- II. Kwak, Tae-Hwan, 1938-

JZ6009.K7.N67 2007
327.1'747095193--dc22

2007013130

ISBN 978-0-7546-7176-3

Printed and bound in Great Britain by MPG Books Ltd, Bodmin, Cornwall.

Contents

Notes on Contributors		*vii*
1	Introduction *Tae-Hwan Kwak and Seung-Ho Joo*	1
2	North Korea's Second Nuclear Crisis and the Six-Party Talks *Tae-Hwan Kwak*	15
3	The Bush Administration and North Korea's Nuclear Policy *Edward A. Olsen*	45
4	U.S. Policy Toward North Korea Under George W. Bush: A Critical Appraisal *Curtis H. Martin*	65
5	China and the North Korean Crisis: The Diplomacy of Great Power Transition *Jeremy Paltiel*	95
6	Japan and the Second North Korean Nuclear Crisis *Yoshinori Kaseda*	111
7	Russia and the North Korean Nuclear Crisis *Seung-Ho Joo*	133
8	U.S.-North Korean Negotiating Behavior and the Six-Party Talks *Scott Snyder*	151
9	Korean Peninsula Energy Development Organization (KEDO) A Bridge Too Far? *C. Kenneth Quinones*	167
10	The Six-Party Talks and Security Cooperation in Northeast Asia *Yong-Sup Han*	191
Index		*207*

Notes on Contributors

Editors

Seung-Ho Joo is Associate Professor of Political Science, at the University of Minnesota-Morris. He received a B.A. in Political Science from Yonsei University and a Ph.D. in Political Science from the Pennsylvania State University. His research interest areas include Russian foreign policy, Russo-Korean relations, and Korean foreign relations. Dr. Joo is the author of *Gorbachev's Foreign Policy toward the Korean Peninsula, 1985-1991: Reform and Policy* (Edwin Mellen, 2000); and co-editor of *The United States and the Korean Peninsula in the 21st Century* (Ashgate, 2006); *The Korean Peace Process and the Four Powers* (Ashgate, 2003); and *Korea in the 21st Century* (Nova, 2001). He has authored over 45 book chapters and journal articles, with the latter appearing in *Pacific Affairs*, *Journal of Northeast Asian Studies*, *American Asian Review*, *Comparative Strategy, Arms Control, Asian Perspective, Pacific Focus,* and *The Korean Journal of Defense Analysis.* He is currently working on a book manuscript on Russia and Korea, 1992-2006. He is Associate Editor for North America of *Pacific Focus* (2003-present) and former President of the Association of Korean Political Studies in North America (2003-05). He was a Korea Foundation Field Research Fellow (2005) and a Humphrey Institute Policy Fellow (1997-98).

Tae-Hwan Kwak is Professor Emeritus at Eastern Kentucky University; visiting Professor of the University of North Korean Studies, Seoul; former President of the Korea Institute for National Unification (KINU) (1999-2000); former Director of the Institute for Far Eastern Studies, Kyungnam University, Seoul (1995-99); and is a specialist on East Asian affairs, international politics and foreign policy. He taught international relations and East Asian politics in 1969-99 at Eastern Kentucky University. He received a B.A. in English from Hankuk University of Foreign Studies, an M.A. in International Relations from Clark University, and a Ph.D. in International Relations from Claremont Graduate University. Dr. Kwak is the author of *In Search for Peace and Unification on the Korean Peninsula* (1986) and *The Korean Peninsula in World Politics* (1999, in Korean). He is editor and co-editor of 25 books, including *The United States and the Korean Peninsula in the 21st Century* (Ashgate, 2006); *The Korean Peace Process and the Four Powers* (Ashgate, 2003); *Korea in the 21st Century* (Nova Science, 2001); *The Major Powers of Northeast Asia: Seeking Peace and Security* (Lynne Rienner, 1996); *The Four Powers and Korean Unification Strategies* (1997); and *The U. S.-ROK Alliance in Transition* (1996). Dr. Kwak was editor-in-chief of *International Journal of Korean Unification Studies* (1999-2000) and editor-in-chief of *Asian Perspective* (1995-99). He is a columnist and a freelance writer for Korean newspapers and monthly magazines.

Contributors

Yong-Sup Han is Director of the Research Institute on National Security Affairs and Professor at the Korea National Defense University. He was Special Assistant to the Korean Minister of Defense (1993) and a Senior Staff Member to the South-North Joint Nuclear Control Commission (1991-92). Dr. Han holds a BA from Seoul National University, an MA in Political Science from Seoul National University, a master's degree in Public Policy from Harvard University, and a Ph.D. in Security Policy from RAND Graduate School. His publications include: Peace and Arms Control on the Korean Peninsula, Sunshine in Korea (coauthored with Norman Levin); "Time for Conventional Arms Control on the Korean Peninsula," Arms Control Today (December 2000); "North Korean Behavior in Nuclear Negotiations," *The Nonproliferation Review* (Spring 2000); and *Nuclear Disarmament and Nonproliferation in Northeast Asia* (The United Nations, 1995).

Yoshinori Kaseda is Associate Professor of Politics, University of Kitakyushu, Japan. He has taught at Northern Illinois University and Miyazaki International College in Japan. He received his B.A in philosophy from the University of Kumamoto in Japan in 1993 and his Ph.D. in political science from NIU in 2006. He has published articles in journals such as *World Affairs, International Journal of Korean Unification Studies,* and *Journal of Political and Military Sociology.* He has also contributed chapters to *Korea in the 21st Century* (Nova Science, 2001) and *The Korean Peace Process and the Four Powers* (Ashgate, 2003).

Curtis H. Martin is Professor of Political Science at Merrimack College in North Andover, Massachusetts. He holds the Ph.D. from Tufts Fletcher School. His teaching areas are international politics, comparative politics and foreign policy, and United States foreign policy. His primary research interests are nonproliferation diplomacy, positive and negative sanctions, and case writing. Publications include: "'Good Cop/bad Cop' as a Model for Nonproliferation Diplomacy toward North Korea and Iran," *Nonproliferation Review* (forthcoming); "Going to the United Nations: George W. Bush and Iraq" (Institute for the Study of Diplomacy, Georgetown University); "The Sinking of the *Ehime Maru*: Interaction of Culture, Interests and Domestic Politics in an Alliance Crisis," *Japanese Journal of Political Science,* 5 (2), 2005; "Rewarding North Korea: Theoretical Perspectives on the 1994 Agreed Framework," *Journal of Peace Research,* 39 (1), 2002; "Negotiating with Adversaries after the Cold War: Incentives-Based Diplomacy in United States-North Korean Relations," in A. Cooper Drury and Steve Chan, eds., *Sanctions as Economic Statecraft* (MacMillan, 2001); "Lessons of the Agreed Framework for Using Engagement as a Nonproliferation Tool," *The Nonproliferation Review*, 6 (4) (Fall 1999); and Martin and B. Stronach, *Politics East and West: Political Culture in Japan and Britain* (M.E. Sharpe, 1991).

Edward A. Olsen is Professor of National Security Affairs & Asian Studies at the Naval Postgraduate School, Monterey, California. He has been on the NPS faculty since 1980. Prior to that (1975-80) he was a political analyst on Japan and Korea at the U.S. Department of State, Bureau of Intelligence and Research (INR), Office of East Asian Affairs. He has a B.A. in History (UCLA, 1968), M.A. in East Asian Studies (UC Berkeley, 1970), Certificate in Japanese language (Stanford University Inter-University Center in Tokyo, 1969-70), and Ph.D. in International Studies (The American University, School of International Service, 1974). He has numerous publications in Asian politics, security, and U.S. foreign/defense policy. He is the author of 8 books, 7 monographs, and was an editor for 4 books. His most recent books are: *U.S. National Defense for the Twenty First Century: The Grand Exit Strategy* (Frank Cass/Taylor & Francis, 2002); *Toward Normalizing U.S.-Korea Relations: In Due Course?* (Lynne Rienner, 2002); and *Korea: A Nation Divided* (Praeger Security International, 2005). The Rienner book also was published in Korean as *Hanmi kwangae ui sae jipyung* [New Horizons of US-Korea Relations] (Ingan Sarang, 2003). His latest monograph is *Homeland Security: Learning from Japan* (The Independent Institute, 2005). He has authored 68 book chapters, 142 journal articles, and 93 Op-Ed columns in major U.S. and Asian media. He currently is working on another book: *Reforming U.S. Foreign Policy* (due in 2008).

Jeremy Paltiel is Associate Professor in Political Science at Carleton University in Ottawa, Canada. He has published on diverse topics on China including civil-military relations, China's accession to the WTO, Communist Party politics, and the Chinese tradition and human rights. Recent publications include "Does a Half-full Glass Justify a Leap of Faith?: Incremental Change and Human Rights in China," *International Journal,* 61 (2) (spring 2006). His manuscript entitled *The Empire's New Cloths: Cultural Particularism and Universal Value in China's Quest for Global Status* on the symbolic importance and implications of sovereignty for China's political evolution is in press with Palgrave. Other upcoming book chapters include, "Hinges and Latches on the Open Door: The Normative Parameters of China's WTO Accession," David Zweig, ed., *International Political Economy and China's Reforms* (Routledge, 2006); "Peaceful Rise? Soft Power? Human Rights in China's New Multilateralism," Guoguang Wu, ed., *China's Diplomacy of Multilateralism* (Routledge, Upcoming).

C. Kenneth Quinones is currently the Director of Global Studies and Professor of Korean Studies at Akita International University in Japan. He has been involved with Northeast Asia since 1962 as a soldier, scholar, and diplomat. He has lived and worked in South and North Korea, plus Japan and visits China often. As a U.S. diplomat, he witnessed South Korea's democratization during the 1980s and was directly involved in North Korea's opening to the outside world during the 1990s. He was the first U.S. diplomat to visit North Korea, was a member of the US negotiating team that resolved the first Korean nuclear crisis, and served as the State Department's liaison officer while living at the Yongbyon Nuclear Research Center and in Pyongyang for nine months in 1995 and 1996. After retiring from the U.S. Foreign Service in 1997, he arranged U.S.-North Korea educational and agricultural exchanges for the Asia Foundation and Mercy Corps. Between 2001 and 2005, he concentrated on writing and commenting about U.S. relations with Northeast Asia, particularly the Korean Peninsula. Since 2000, he has published three books: *The*

North Korea Nuclear Crisis—Off the Record Memories, (translated into Korean and Japanese in 2000); *Beyond Diplomacy: Implementation of the Agreed Frame Work* (published in Japanese in 2003); and *The Complete Idiot's Guide to Understanding North Korea* (published in English by Penguin Publishers' Alpha Books in 2004). His monthly columns appear in Japan's *Mainichi Shimbun* and *Sanyo Shimbun*. He has contributed numerous scholarly studies and articles to academic journals, edited books and newspapers in the U.S., South Korea, and Japan. He has had numerous interviews with North American, East Asian and European radio and televisions stations. Dr. Quinones holds a Ph.D. in History and East Asian Languages from Harvard University.

Scott Snyder is a Senior Associate in the International Relations program of The Asia Foundation and Pacific Forum CSIS, and is based in Washington, DC. He spent four years in Seoul as Korea Representative of The Asia Foundation during 2000-2004. Previously, he served as a Program officer in the Research and Studies Program of the U.S. Institute of Peace, and as Acting Director of The Asia Society's Contemporary Affairs Program. His publications include *Paved with Good Intentions: The NGO Experience in North Korea* (2003) (co-edited with L. Gordon Flake) and *Negotiating on the Edge: North Korean Negotiating Behavior* (1999). Snyder received his B.A. from Rice University and an M.A. from the Regional Studies East Asia Program at Harvard University. During 2005-06, he was a Pantech Visiting Research Fellow at Stanford University's Shorenstein Asia Pacific Research Center (APARC). He was the recipient of an Abe Fellowship during 1998-99, and was a Thomas G. Watson Fellow at Yonsei University in South Korea in 1987-88.

Chapter 1

Introduction

Tae-Hwan Kwak and Seung-Ho Joo

North Korea's test explosion of a nuclear bomb on October 9, 2006 sent a shockwave throughout the world and totally changed the strategic equation in the Korean peninsula and Northeast Asia. With that, the Democratic People's Republic of Korea (DPRK or North Korea) declared itself a nuclear power, and the international community led by the US scrambled to impose sanctions against North Korea.

The second North Korean nuclear crisis began in October 2002 when North Korea allegedly admitted a secret nuclear program based on highly enriched uranium (HEU). Shortly thereafter, the 1994 U.S.-DPRK Agreed Framework quickly dissolved and North Korea accelerated its nuclear weapons program without external constraints. The six-party talks, involving the DPRK, the Republic of Korea (ROK or South Korea), the U.S., China, Japan, and Russia, convened in Summer 2003 to peacefully resolve the crisis but failed to make a breakthrough agreement. Pyongyang officially announced in February 2005 that it had nuclear weapons, test-fired seven ballistic missiles (including a Taepodong-II long-range missile) in July 2006, and tested its first nuclear device underground in October 2006. With the North Korean nuclear test, the unresolved nuclear crisis entered into a precarious, perilous phase. The North Korean test explosion of a nuclear device was to have serious and far-reaching implications for inter-Korean relations, Northeast Asian security, and the U.S. global war on terrorism; Korea's peaceful unification seemed remote, Northeast Asian security became unpredictable, and the U.S. war on terrorism faced new challenges.

This book is designed to present in-depth analyses of the inter-Korean and international dynamics of North Korea's second nuclear crisis, provide new insights into the on-going six-party talks, and offer some creative formulas to resolve the nuclear crisis through peaceful, diplomatic means. This volume specifically delves into the following questions: Why did the second North Korean nuclear crisis begin and how has it evolved?; How has North Korea's nuclear issue affected inter-Korean reconciliation (or lack thereof) and vice versa?; What are the interests and policies of the major powers, that is, the U.S., China and Japan, at the six-party negotiating table?; What are the short-term and long-term implications of North Korea's nuclear blast?; and How should the two Koreas and the surround powers untangle North Korea's nuclear knot for a lasting peace and security in and around the Korean peninsula?

A Brief Historical Overview

The six-party nuclear talks were held on-again, off-again since Summer 2003 but failed to resolve the North Korean nuclear crisis. Hostile policies, inflexible attitudes, mutual distrust, and incompatible positions of North Korea and the U.S. led to the stalemate at the six-nation negotiating table. North Korea has demanded direct negotiations with the U.S. on the nuclear issue and insisted that the U.S. make appropriate compensations for abandoning its nuclear programs. The George W. Bush administration, however, preferred a multilateral framework for resolving North Korea's nuclear issue and insisted on complete, verifiable and irreversible dismantlement (CVID) of North Korean nuclear programs. Although main differences among the six participants, particularly between the U.S. and the DPRK, remained unresolved, the talks proved useful. The six nations had substantive discussions about North Korea's nuclear problem. The U.S. and the DPRK showed some flexibility on certain issues, but remained stubborn on key issues such as North Korea's suspected HEU program and CVID of North Korea's nuclear programs.

The six-party talks made substantial progress on September 19, 2005 when the parties produced a Joint Statement of Principles for establishing the verifiable denuclearization of the Korean peninsula. Although the terms of the accord were broad and general, the Joint Statement marked the first concrete agreement among the six parties. It was designed to serve as the basis for further talks on the timing of the North Korea's dismantlement of its nuclear weapons programs and the corresponding provision of economic aid, diplomatic relations, and other incentives for the DPRK. The U.S., however, imposed financial sanctions against North Korea in the same month, and North Korea refused to return to the six-party negotiating table to devise a roadmap for implementing the 9.19 Joint Statement. The DPRK maintained that it would not return to the six-party talks unless the U.S. lifted the financial restrictions first. The U.S., in contrast, held that the financial sanctions were imposed to prevent money laundering and other illegal activities and should be separated from the nuclear talks. The U.S. financial sanctions and North Korea's stubbornness regarding the sanctions proved the major obstacle to the resumption of the six-party talks.

North Korea escalated the crisis by test-firing seven ballistic missiles on July 5, 2006. Among the missiles test-fired was a long-range Taepodong-II, which is potentially capable of reaching the U.S. west coast. This move invited immediate, negative reactions from the international community. The UN Security Council adopted Resolution 1695 on July 15, condemning the tests and paving the way for other countries to impose financial sanctions against the DPRK. Additionally, South Korea and Japan unilaterally took punitive actions against North Korea. Pyongyang's defiant missile tests mounted pressure from all other nations of the six-party talks and produced a high level of tension and anxiety between the DPRK and other state parties. North Korea may have carried out the missile tests in order to increase its leverage at the six-party negotiations. After concluding that the six-party negotiating process was futile and counter-productive, Kim Jong-Il may have wanted to vent out his frustrations and at the same time demonstrate the formidable military capability of his country.

The DPRK announced on October 9, 2006 it had successfully conducted its first test of a nuclear weapon. The historic event shocked the world. It appears that Chairman Kim Jong-Il calculated that a nuclear-armed North Korea could better induce the U.S. to direct bilateral negotiations. The North's nuclear test promptly led to the adoption of the UN Security Council Resolution 1718 under Chapter 7, Article 41 of the UN Charter aimed at punishing North Korea for its nuclear test through economic sanctions. The DPRK rejected the resolution and threatened "physical countermeasures" against any state that tried to enforce the UN sanctions. The North Korean nuclear test failed to push the U.S. into bilateral talks but instead put more pressures on Pyongyang through the UN Security Council. It also posed a grave security threat to the Korean peninsula and Northeast Asia. As a result of China's active, skillful mediation, the DPRK announced on November 1, 2006 that it would return to the six-party talks if the U.S. would discuss and resolve the financial sanctions issue within the framework of the six-party talks.

The Bush administration's hard-line policy toward North Korea proved counter-productive, and North Korea's missile and nuclear tests were a testimony to the limits of the U.S. hostile approach to Pyongyang. Additional pressures on North Korea were likely to be ineffective since China and the ROK refused to push North Korea into the corner. China has argued that the problem of financial sanctions must be resolved for the denuclearization of the Korean peninsula and the U.S. should engage in bilateral talks with the North for the six-party talks' resumption. Russia shared the international community's deep disappointment and anger over North Korea's missile and nuclear tests, but did not want to push for Pyongyang's regime change or collapse. South Korea reluctantly began to apply limited economic sanctions against the North for the first time in the wake of the missile tests, but still refused to go along with far-reaching, comprehensive sanctions against its communist neighbor. Like the U.S., Japan maintained a hard-line stance toward the North. Japan was already taking measures to incrementally increase sanctions in accordance with the UN Security Council resolutions. The U.S. and Japan were seeking to apply the Proliferation Security Initiative (PSI) to North Korea. PSI is a policy of the U.S. as part of its global war on terror. If implemented, it will allow states to search and interdict North Korean ships and aircraft for contraband items (the weapons of mass destruction, missiles, illegal weapons, and others). China, Russia, and the ROK have resisted to full participation in PSI.

Following the missile and nuclear tests, inter-Korean relations sunk to the lowest point since the inter-Korean summit of June 2000. After the missile tests, Seoul suspended its regular food and fertilizer aid to the North. The nuclear test jeopardized inter-Korean economic projects such as the Gaesong Industrial Complex and the Mount Geumgang tourism resort. Furthermore, the nuclear test compelled the ROK to re-evaluate its engagement policy of national reconciliation and cooperation with the DPRK. The debate over the engagement policy divided and fragmented South Korean society.

North Korea's nuclear test had multiple implications. In the short-term, the test appeared to weaken the China-DPRK alliance, but strengthen U.S.-China cooperation and the U.S.-ROK alliance. Its long-term impact on the Northeast Asian security was to be manifold. First, it would pose a major challenge to the nuclear non-proliferation regime and might spark a chain-reaction of nuclear proliferation in Northeast Asia and around the world. For now, Japan and the ROK may settle with the extended nuclear deterrence offered by the U.S. but, in the long-run, they might want to acquire nuclear deterrence of their own. Taiwan would be tempted to go nuclear as well. Iran might follow in the footsteps of North Korea, accelerating and securing nuclear weapons for itself. Should this happen, Turkey and Saudi Arabia would be likely to follow suit. Second, the North's nuclear test certainly prompted Tokyo to accelerate a missile-defense (MD) system in cooperation with Washington, which would in turn push Beijing to spend more money developing advanced nuclear weapons. Thus, a new nuclear arms race among Japan, China and other countries of Northeast Asia might ensure. Third, North Korea might sell nuclear weapons to rogue states and terrorist groups. President George W. Bush said, "The transfer of nuclear weapons or material by North Korea to states or non-state entities would be considered a grave threat to the United States, and we would hold North Korea fully accountable of the consequences of such action."[1] If the North would sell a nuclear weapon to a terrorist organization like Al Qaeda that could use it against the U.S. or one of its allies, then the U.S. would very likely retaliate militarily against Pyongyang.

The Nuclear Agreement (February 13, 2007)

The third session of the fifth round of the six-party talks, convened on February 8-13, 2007, produced a landmark agreement titled "Initial Actions for the Implementation of the Joint Statement." The 16 hour marathon-like, all-night negotiating session finally led to a compromise. In the accord, the six nations agreed to take the initial actions as the first step toward the implementation of the September 19, 2005 Joint Statement in a phased manner in line with the principle of "action for action."

The Gist of the Agreement

The following are key points of the February 13 agreement.[2] This action plan has two phases—the initial phase and the disablement phase— to implement the September 2005 joint agreement.

First, at the initial phase, the DPRK must shut down and seal its main nuclear facilities at Yongbyon within 60 days. IAEA inspectors should be allowed to monitor and verify the process. In return, North Korea will get energy, food, and other aid worth 50,000 tons of heavy fuel oil (HFO). Second, at the disablement phase, the DPRK must provide a complete list of its nuclear programs and disable all existing nuclear facilities. In return, the DPRK will get aid in corresponding steps worth 950,000 tons of heavy fuel oil or the equivalent in the form of economic or humanitarian

1 "World Aligns against N. Korea for Nuclear Test," *Associated Press*, October 9, 2006.

2 The full text of the February 13 Agreement is available online at <http://americancorners.or.kr/e-infousa/wwwh5668.html>

aid, from China, the United States, the ROK, and Russia—details of which will be addressed in later working group discussions. It would be worth around $300 million at current prices for Asian benchmark high-sulfur heavy fuel oil, which is used in power stations, shipping, and elsewhere. Third, the United States and the DPRK will begin bilateral talks to normalize their diplomatic relations and the U.S. will begin the processes of removing North Korea from its designation as a terror-sponsoring state and also ending U.S. trade sanctions against the DPRK. However, no deadline has been set. Fourth, Japan will begin bilateral talks with the DPRK to normalize their relations in accordance with the Pyongyang Declaration, an agreement made between the two nations on September 17, 2002. Fifth, after 60 days, foreign ministers of all six nations will meet to confirm the implementation of the joint agreement and discuss security cooperation in Northeast Asia. The directly related parties, that is, the U.S., China, and two Koreas, will hold a separate forum on negotiations for a permanent peace regime to replace the 1953 Korean armistice agreement. Sixth, five working groups will be established to carry out the initial actions and implement the joint statement: (1) denuclearization of the Korean peninsula chaired by China, (2) normalization of U.S.-DPRK relations, (3) normalization of Japan-DPRK relations, (4) economic and energy cooperation chaired by the ROK and (5) Northeast Asia peace and security cooperation chaired by Russia. Seventh, the sixth round of the six-party talks will be held on March 19, 2007 to discuss actions to be taken for the next phase.

In short, at the initial phase, the six parties agreed on the beginning steps toward eventual dismantling of North Korea's nuclear arms programs. The deal requires the DPRK to shut down its Yongbyon reactor within 60 days in exchange for 50,000 tons of fuel oil or equivalent aid. After the 60-day period, the DPRK will receive another 950,000 tons of fuel oil, or equivalent aid at the disablement phase, when it takes further steps to disable its nuclear facilities.

Bush Administration's New Flexibility

The nuclear deal on the initial actions was made possible largely because President Bush was willing to take a new more flexible approach to reach an agreement with North Korea.[3] Ever since North Korea's second nuclear crisis started in 2002, the Bush administration had insisted that North Korea should not be rewarded for its "bad behavior" and that it should first give up its nuclear ambitions before it could receive anything in return. After the November 2006 mid-term elections, hard-liners in the Bush administration, such as former Defense Secretary Donald H. Rumsfeld and former Ambassador to the United Nations John R. Bolton, who had opposed concessions to North Korea left the administration. As a result, the hard-liners' position considerably weakened and the pragmatists' stature in the Bush administration enhanced.

3 Glenn Kessler and Edward Cody, "U.S. Flexibility Credited in Nuclear Deal with N. Korea," *Washington Post,* February 14, 2007.

The decision-making process of the Bush administration in the case of the February 13 nuclear deal differed from the usual procedures. Secretary of State Condoleezza Rice bypassed a usual policy review process to get approval of a deal with North Korea from President Bush. Four key members in the Bush administration—Secretary Rice, Assistant Secretary of State for East Asian and Pacific Affairs Christopher Hill, National Security Adviser Stephen J. Hadley, and President Bush—were directly involved in the decision-making process when a deal with North Korea was vetted after Hill and his North Korean counterpart Kim Kye-gwan engaged in negotiations in Berlin in January 2007. The usual procedures in the Bush administration were to review "the details though an interagency process that ordinarily would have brought in Vice President Dick Cheney's office, the Defense Department and aides at the White House and other agencies who had previously objected to rewarding North Korea before it gives up its weapons." But this time the usual procedures were curtailed.[4]

Fierce attack on the deal came from neo-conservatives. John R. Bolton bluntly stated that it was a "bad deal" and Undersecretary of State for Arms Control and Disarmament Robert Joseph criticized the deal, stating, "the new agreement was no better, and perhaps worse, than one signed by President Clinton in 1994."[5] President Bush and Secretary Rice wanted a deal, because "as the Bush presidency enters its final years, two foreign policy goals have risen higher on the agenda: peace between the Israelis and Palestinians and disarming North Korea."[6] President Bush approved a deal with North Korea and Hill signed it in Berlin in January 2007, which led to the February 13 nuclear agreement in Beijing.

The Bush administration has shown a new flexibility in dealing with North Korea. The Berlin bilateral talks between the U.S. and the DPRK were personally approved by President Bush although he had previously insisted for four years that he would not allow direct bilateral negotiations. Bush's decision to change his policy toward the North is praiseworthy and realistic, and it was overdue. Furthermore, it was the key to the breakthrough accord in February. He no longer insisted that the DPRK must abandon its nuclear programs first, but made concessions by accepting North Korea's original position—suspension of its Yongbyon nuclear facilities—and by unfreezing some of the North Korean bank accounts at the Banco Dalta Asia (BDA) in Macao. Assistant Secretary Hill told the DPRK and China on February 14 that the U.S. would resolve a dispute over the BDA issue within 30 days. This clearly indicated a significant change from the Bush administration's previous position that the BDA dispute was a law-enforcement issue that should be separate from the nuclear negotiations. The change certainly was a big step in the right direction, and the only regret is that he did not make such a decision during his first-term of presidency.

President Bush praised the deal, calling it a "first step" toward implementing the September 19, 2005 statement. Secretary of State Rice said the deal was part

4 David Sanger and Thom Shanker, "Rice Is Said to Have Speeded North Korea Deal," *New York Times*, February 16, 2007.

5 Ibid.

6 Ibid.

of a "broad and comprehensive effort," not only to achieve a nuclear-free Korean Peninsula, but also "to advance the future of peace and prosperity in Northeast Asia."[7] However, the new deal was not open-ended. The DPRK will get no more than the one-time emergency energy supply equivalent to 50,000 tons of heavy fuel oil, worth about $12 million, unless it takes further action. This accord would be carried out through the six-party process. As the Washington Post editorial pointed out, "it is wrong to argue that the administration has simply reverted to the Clinton-era arrangement that it repudiated in 2002, and if it is rewarding North Korea's misbehavior, the bribe is a small one."[8]

The U.S.-China-ROK trilateral cooperation at the third session of the talks was essential to the breakthrough deal. China was so shocked and embarrassed by Pyongyang's nuclear test in October 2006 that it became a driving force in the talks. President Bush, in a conversation on February 15, congratulated Chinese President Hu Jintao about the key role China played in reaching the nuclear deal. Bush told Hu that "it was now up to the leader of North Korea to live up to the commitments made in order to create a better life for the North Korean people."[9] President Hu reiterated his country's commitment to helping implement the agreement, stressing that China was "willing to maintain close communication and cooperation with the United States and other parties concerned ... to play a constructive role" in the denuclearization process. He also said, "A full implementation of the document is not only of great significance for safeguarding peace and stability on the Korean peninsula and in Northeast Asia, but also serves the common interests of all parties concerned."[10]

Key Issues Yet to be Resolved

A few critical, key issues have not been settled and subsequent six-party negotiations should resolve differences over HEU, the disablement of nuclear facilities, and nuclear weapons. First of all, the question remains whether North Korea and the other five participants will abide by the nuclear deal. As Chief U.S. negotiator Hill cautioned, difficult work remains to implement the accord with the DPRK.[11] There will be a long, hard road ahead. Shutdown, sealing and disabling of North Korea's nuclear facilities will not necessarily lead to a complete, verifiable, and irreversible dismantlement of all nuclear programs, including the declared nuclear weapons, which will be a more difficult and long-term issue yet to be negotiated.

The North Korean Central News Agency reported, perhaps intentionally, that the agreement required only a temporary suspension of North Korea's nuclear facilities,

---

7 "Secretary Rice's Briefing on the Agreement Reached at the Six Party Talks in Beijing," U.S. Department of State, February 13, 2007 <http://americancorners.or.kr/e-infousa/wwwh5667.html>

8 "Nuclear Bargaining," *Washington Post,* February 14, 2007: A18.

9 Steve Holland, "Bush seeks to tamp down revolt on North Korea deal," *Reuters*, February 15, 2007.

10 "China Intent on Making Nuke Deal Happen," *Associated Press*, February 16, 2007.

11 Linsay Beck, "Hard work yet to come on North Korea nuclear deal," *Reuters,* February 13, 2007.

"At the talks the parties decided to offer economic and energy aid equivalent to one million tons of heavy fuel oil in connection with the DPRK's temporary suspension of the operation of its nuclear facilities."[12] This statement deliberately misled the North Korean people. Hill dismissed this report saying, "they [North Koreans] do it for various domestic reasons." He insisted that if the North Koreans cheated on the agreement, "we will know that pretty soon" and they would be reneging on a commitment to China and four other countries, not just the United States.[13]

In the meantime, while in the airport upon his return to Pyongyang from Beijing on February 15, 2007, the DPRK Vice Foreign Minister Kim Kye-gwan told the Russian ambassador to Pyongyang and a senior diplomat from the Chinese embassy that the DPRK was ready to implement the February 13 agreement on the initial steps toward denuclearization, according to the Kyodo News agency.[14]

Second, one weakness of the landmark deal is that there was no mention about North Korea's declared nuclear weapons, estimated to be as many as 10 bombs from a stockpile of perhaps 50 kilograms of plutonium it has produced. This key issue remains unresolved. Whether Chairman Kim Jong-Il will abandon his nuclear bombs and plutonium will be tested in the near future. According to the action plan in the deal, at the next phase, the DPRK must permanently disable the Yongbyon facilities and provide a "complete declaration of all nuclear programs" in exchange for the equivalent of 950,000 tons of oil. How and when it will accomplish the disablement, how it will be verified, and what else the DPRK may receive in return remain to be worked out by the working groups. The HEU project, which the DPRK has denied, is another vexing issue yet to be resolved. Kyodo News reported that the U.S. initially wanted to include North Korea's abandonment of uranium enrichment in a draft agreement proposed by China at the six-party talks, but agreed to drop it after North Korea rejected the idea.[15] The uranium issue is expected to resurface during the second phase. The six-party nuclear talks will eventually have to address the two issues of North Korea's nuclear arsenal and its HEU project.

Third, another key issue to be observed by the DPRK during the initial phase of action is to meet the disclosure requirements. The nuclear disarmament deal requires the DPRK to submit a report of all of its nuclear programs, including uranium-based ones, to the International Atomic Energy Agency (IAEA). Will the DPRK sincerely meet the disclosure requirements? Will it honestly disclose the amount of plutonium it has produced? Will it admit to the HEU program? How will its disclosure be verified? These are just a few of the many questions to be addressed at the working group meetings. Despite all the weaknesses, this accord was a first step toward the North Korean denuclearization.

---

12 "Third Phase of Fifth Round of Six-Party Talks Held," *KCNA*, February 13, 2007 <http://kcna.co.jp/index-e.htm>

13 Steve Holland, "Bush Seeks to Tamp Down Revolt on North Korea Deal," *Reuters*, February 15, 2007.

14 *Kyodo News*, February 15, 2007; "Report: North Korea Ready to Disarm," *Associate Press*, February 15, 2007.

15 "Gave in to N. Korea over uranium enrichment plan: source," *Kyodo News*, February 19, 2007.

It is possible that Chairman Kim Jong-Il made a strategic, bold decision to give up nuclear ambitions, in view of a long-term perspective and of North Korea's domestic and international problems. His prudent decision will surely guarantee his regime's security and survival and will eventually contribute to the Korean peace-making process and North Korea's economic prosperity. We do agree with the argument made by Robert Carlin and John W. Lewis that "Denuclearization, if still achievable, can come only when North Korea sees its strategic problem solved, and that, in its view, can happen only when relations with the United States improve."[16] The initial action plan in the agreement is expected to pave the way to the U.S.-DPRK normalization of diplomatic relations.

Contributions to the Book

As discussed above, the North Korean second nuclear crisis is one of the most intractable and complex issues in today's world. The North Korean nuclear crisis, if unresolved, will have profound negative effects on the future of the Korean peninsula and Northeast Asian security. Thus, a clear understanding and objective analysis of the second North Korean nuclear crisis and a careful examination of the most desirable, feasible ways of resolving it are imperative. It is with this goal in mind that this volume was conceived and organized. This book consists of ten chapters, with topics ranging from North Korea's second nuclear crisis, the Bush administration's policy on North Korea, major powers' interests and policies towards North Korea's nuclear issue, North Korean negotiation behavior, the KEDO (Korean Peninsula Energy Development Organization), the provision of light-water nuclear reactors (LWRs) to North Korea, and North Korea's nuclear test explosion and its security implications. The contributing authors are distinguished specialists and experts in the field of U.S.-Korean relations and Northeast Asian security, who will offer valuable expertise and insights into the dynamics of North Korea's nuclear crisis and Northeast Asian security. To help the reader better understand the gist of the book, a synopsis of each chapter is offered below.

Chapter 2, by Tae-Hwan Kwak, "North Korea's Second Nuclear Crisis and the Six-Party Talks," reviews and evaluates the five rounds of the six-party talks as a multilateral framework for resolving North Korea's nuclear issue. The author argues that the six-party process is the best means to resolve the North Korean nuclear crisis and maintains that direct bilateral talks between the U.S. and North Korea are essential to a peaceful and diplomatic resolution of North Korea's nuclear standoff. He urges both the U.S. and North Korea to engage in direct negotiations without preconditions within the six party framework and to become flexible about their respective positions, keeping in view reaching a solution through compromise. This chapter has three specific goals: (1) to evaluate the fourth and fifth rounds of the six-party talks; (2) to examine the U.S. financial sanctions against North Korea and concerned parties' positions on the sanctions; and (3) to make policy

16 Robert Carlin and John W. Lewis, "What North Korea Really Want," *Washington Post*, January 27, 2007, A19.

recommendations for implementing the September 19 joint statement to realize a denuclearized Korean peninsula.

Chapter 3, by Edward A. Olsen, "The Bush Administration and North Korea's Nuclear Policy," evaluates George W. Bush's North Korea policy from a critical perspective and recommends a new North Korea policy relying on soft power. U.S. relations with North Korea have ranged from poor to hostile – never positive. The tensions surrounding North Korea's nuclear weapons agenda have exacerbated the level of hostility and put pressure on Washington to create a more effective approach to dealing with North Korea. Those tensions and pressures were intensified in the post-9/11 geopolitical atmosphere, leading the Bush administration to pursue hawkish goals regarding the DPRK, which collided with South Korea's advocates of an engagement approach within inter-Korean relations. In order to understand the factors shaping these relationships it is necessary to grasp the dynamics which characterize the evolution of U.S. relations with both Koreas during and since the Korean War. They are evaluated in this analysis. Against this background it is evident that the Bush administration's policies toward North Korea have not been very effective. There is a pressing need for greater coordination of the United States' policies with South Korea, China, and Japan. In particular, the United States should reject its negative "no" approach to the Kim Jong-Il government in North Korea in favor of adapting a more positive no/Roh (South Korean President Roh's name is pronounced "no") approach within U.S. foreign and defense policies. In doing so the United States should incorporate Japan's "comprehensive security" paradigm based on using soft power as a way to engender reciprocal confidence and set the stage for creating a Korean peace treaty and greater attention to post-war inter-Korean reconciliation and reunification.

Chapter 4, by Curtis H. Martin, "U.S. Policy Toward North Korea under George W. Bush: A Critical Appraisal," evaluates, from the perspective of pertinent system, state, and individual level theory, the unfolding of the Bush administration's policy for addressing North Korea's nuclear programs. It explores the question of whether that policy has been "effective harmony of different goals," as the administration claims, or something more akin to "ostrich engagement," "drift" or "hostile neglect" as claimed by critics. Since its inception, the Bush administration's policy has been grounded in a strong—though not universally held—preference for "asphyxiation" rather than "oxygen," for pressure and isolation rather than rewards. That policy has been constrained, but never transformed, by system and state level realities that have driven the United States to eschew forceful unilateral action and to pursue its objectives through multilateral regional diplomacy. Throughout, however, the strong preference of core decision makers for isolation has reined in perceived tendencies to "reward bad behavior" pending a more propitious environment for a policy of pressure. This preference has led U.S. policy to exhibit characteristics of "hawk engagement" and "temporary appeasement," in which negotiations assume primarily tactical significance. Checked in its ability to pursue an energetic and forward leaning policy of pressure, and unwilling wholeheartedly to pursue a negotiated solution, U.S. policy has appeared to waver between the two strategies while lacking the advantages of either. The opportunity to pursue the preferred pressure policies arrived with implementation of new PSI-like financial sanctions in Bush's second term and the shift in the geopolitical terrain in the wake of North Korea's missile tests. By the end of 2006, US policy toward the DPRK still faced daunting constraints, but these were fewer and less decisive than they had been from 2002-05, breathing new life into the administration's default policy.

Chapter 5, by Jeremy Paltiel, "China and the North Korean Crisis: The Diplomacy of Great Power Transition," examines China's relationship to the North Korean nuclear crisis as a microcosm of its emergent global role. The crisis places China's relationship with a traditional alliance with a socialist comrade in arms, the DPRK, in balance with an emergent regional leadership role focused on its closest trading partners in Northeast Asia, Japan and the ROK. Against the background of this regional role, China is seeking to seize the initiative in defusing a potential global nuclear proliferation crisis from the world's sole superpower, the United States. In hosting the six-party talks, and especially through brokering the September 19, 2005 Statement, China has striven to put into practice its "new security concept"; the resolution of conflict through dialog aimed at producing "win-win" agreements. To achieve this strategic goal, China has had to contend with a North Korean strategy premised on brinkmanship and a U.S. administration that is rhetorically pledged to use multilateral pressure to get a charter member of the eponymous "axis of evil" to dismantle its nuclear program verifiably and unconditionally. Nuclear diplomacy on the Korean peninsula tests the limits of China's capacity to provide global "public goods" in its quest to become a responsible great power. The chapter asks whether an inconclusive diplomatic result can boost China's status as a global and regional power.

Chapter 6, by Yoshinori Kaseda, "Japan and the Second North Korean Nuclear Crisis," critically evaluates Japan's policy toward the DPRK including its tough stance at the six-party talks, identifies its problems, and considers what kind of policy stance Japan should take in order to realize a peaceful resolution of the nuclear problem in the near future. After the Cold War, Japan has come to regard the DPRK as the primary military threat to its national security, due to the DPRK's development of the weapons of mass destruction. Arguably, Japan is most strongly threatened by the DPRK's development of nuclear weapon and ballistic missiles, given the present range of the ballistic missiles and the hostile relations between the two countries. Thus, Japan has a very good reason to make an active response to the military threat posed by the DPRK and make a serious effort at resolving the problem of North Korea's nuclear development. In fact, Japan has taken an active military response, enhancing its military capability and strengthening its alliance with the United States. However, its role in resolving the nuclear problem seems limited. At the multilateral talks, Japan has taken a tough, rather uncompromising stance toward the DPRK. It has demanded a comprehensive resolution of the nuclear and missile issues as well as the issue of the DPRK's abduction of Japanese citizens, and made its resolution a precondition for its diplomatic normalization with the DPRK. Japan has also maintained that it would not give economic assistance to the DPRK before the normalization.

Chapter 7, by Seung-Ho Joo, "Russia and the North Korean Nuclear Crisis," examines the Russian policy on North Korea's nuclear issue, evaluates Russia's role in the six-party negotiations for the denuclearization of the Korean peninsula, and

Russian initiatives for peace, security, and prosperity in Korea and Northeast Asia. President Vladimir Putin has pushed for numerous peace initiatives in the Korean peninsula in a bid to safeguard its security and economic interests and enhance its prestige as a major player in the region. At the six-party talks, Russia has sought to resolve North Korea's nuclear crisis peacefully. However, Russia's role in the Korean peace process in general, and at the six-party talks in particular, has been indirect and marginal. Russia's impact on the talks was limited to the format and the procedural matters, and its impact on the substance of the negotiated outcomes has been negligible. The outcome of the six-party negotiations depended largely on the U.S. and the DPRK and the diplomatic skills of China to bridge the gap between them. Russia, however, is cautiously pushing for its security and economic agenda in close coordination with China and it is relying on its growing economy and military capability.

Chapter 8, by Scott Snyder, "U.S.-North Korean Negotiating Behavior and the Six-Party Talks," examines possible lessons learned from the first North Korean nuclear crisis, reviews the progress of the six-party process, and analyzes the tactics and strategies of the U.S. and the DPRK, respectively, as well as the other parties to the talks. The experience of U.S.-DPRK negotiations during the first North Korean nuclear crisis has framed American and North Korean strategies and tactics as they have approached the second crisis. The six-party talks negotiating format itself has been favored by the Bush administration based on its own perception of lessons learned from the Clinton administration's experience. For its part, the DPRK leadership has relied on a range of familiar tactics in the initial rounds of talks, including calculated efforts to utilize brinkmanship and crisis escalation to catalyze a direct response from the Bush administration. In the case of both Washington and Pyongyang, the use of these tactics reveal that until North Korea's nuclear test on October 9, 2006, neither side had truly committed itself to resolving the nuclear issue through a give-and-take negotiation process. Instead, both sides have attempted to shape the environment to their own ends as a vehicle for reaping benefits through the mechanism of negotiations. The North Korean nuclear test achieved Pyongyang's tactical goal of drawing the United States into direct talks, but against a strategic context that initially appeared to be relatively unfavorable to North Korea's longer-term strategic aims.

Chapter 9, by C. Kenneth Quinones, "Korean Peninsula Energy Development Organization (KEDO): A Bridge Too Far?" assesses KEDO's accomplishments and shortcomings as well as some of the key reasons for its demise. In an unprecedented effort to halt nuclear proliferation on the Korean Peninsula and elsewhere, the United States, Japan and South Korea formed an international consortium in 1995 to implement selected aspects of the 1994 U.S.-North Korea Agreed Framework. The accord promised North Korea that the international community would provide it heavy fuel oil to generate electricity until two light-water nuclear reactors went on line. In exchange, Pyongyang promised to "freeze" its nuclear program and "suspend" its withdrawal from the Treaty on the Non-proliferation of Nuclear Weapons (NPT). The Korean Peninsula Energy Development Organization (KEDO) was formed to facilitate the oil deliveries and to construct the reactors. After an extremely difficult birth, KEDO finally began operation in earnest in 1997. No sooner had the supply of heavy fuel oil become routine and reactor construction gotten underway by 2001 than the newly installed Bush Administration attacked KEDO as indicative of the Clinton Administration's "appeasement" policy toward North Korea. KEDO subsequently died a slow death and along with it the Agreed Framework.

Chapter 10, by Yong-Sup Han, "The Six-Party Talks and Security Cooperation in Northeast Asia," addresses the question of how to make the six-party talks a success for resolving North Korea's nuclear crisis and promoting multilateral security cooperation among Northeast Asian countries. More specifically, this chapter examines the methods to military tension-reduction and a durable peace on the Korean peninsula. It also explores the ways to promoting multilateral security cooperation in Northeast Asia after resolving North Korea's nuclear issue and reducing tensions in and around the Korean Peninsula. The six-party talks are a regional security dialogue regime that was organized in August 2003 to resolve North Korea's nuclear issue. The September 2005 Joint Statement of the six-party talks raised hopes for the resolution of the North Korean nuclear issue. In fact, the six-party talks are stalling due to differences between Washington and Pyongyang over North Korea's counterfeiting and North Korea's recent nuclear test. Nonetheless, the value of the six-party talks cannot be discounted because they provided the first forum for six countries (South Korea, North Korea, the United States, China, Japan, and Russia) in Northeast Asia to discuss intensely the most security issue—North Korea's nuclear issue. Though broad in its scope and nature, the Joint Statement addressed problems that North Korea and the other five countries have been facing for a long time. In essence, it commits North Korea to abandoning its nuclear weapons and nuclear programs, while the other parties commit themselves to providing security assurances and economic assistances to North Korea. It remains to be seen whether North Korea will abandon its nuclear programs and nuclear weapons completely and verifiably, in spite of its long-held dream of a nuclear weapon state.

Chapter 2

North Korea's Second Nuclear Crisis and the Six-Party Talks

Tae-Hwan Kwak

The United States maintained in October 2002 that the former Assistant Secretary of State James Kelly confirmed, from North Korean officials in Pyongyang, that the Democratic People's Republic of Korea (DPRK or North Korea) had a covert nuclear weapons program using highly enriched uranium. North Korea's alleged admission of a highly enriched uranium program sparked the second nuclear crisis on the Korean peninsula, openly accelerating North Korea's nuclear weapons development program. It is estimated that the DPRK already has six or seven nuclear devices. Furthermore, Pyongyang officially announced that it had nuclear weapons on February 10, 2005 and tested its first nuclear weapon on October 9, 2006.

The North Korean nuclear issue, as the most important obstacle to the Korean peace process, is a serious international matter. It needs to be resolved as soon as possible peacefully through the six-party nuclear talks. The Bush administration insisted that the North Korean nuclear crisis be resolved through the six-party process. The author has analyzed the origin and development of the second nuclear crisis on the Korean peninsula and, elsewhere in detail, has evaluated the first through the third rounds of six-party talks as a multilateral framework for resolving the nuclear issue.[1]

The author argues that the six-party process is the best means to resolve the North Korean nuclear issue. The author maintains that U.S.-DPRK bilateral talks are essential to a peaceful and diplomatic resolution of North Korea's nuclear standoff. Both need to engage in direct negotiations without preconditions through the six-party process. Both the U.S.'s and North Korea's hard-line policies cannot resolve the nuclear issue peacefully. Therefore, both sides need to be flexible about their respective positions and have the political will to make a compromise.

This chapter has three specific goals: (1) to evaluate the fourth and fifth rounds of the six-party talks as a multilateral framework of resolving the nuclear issue, (2) to take a closer look at U.S. financial sanctions against North Korea and the positions of concerned parties, and (3) to make policy recommendations for implementing the 9.19 joint statement to realize a denuclearized Korean peninsula.

---

1 For a detailed analysis of the six-party talks from the first through the third round, see Tae-Hwan Kwak, "The Six-Party Nuclear Talks: An Evaluation and Policy Recommendations," *Pacific Focus*, vol. 19, no. 2 (Fall 2004), pp. 7-55.

Let us first evaluate the fourth round of the six-party talks to resolve the North Korea's nuclear issue. The peaceful resolution of the nuclear issue is a *sine qua non* for building a peace regime on the Korean peninsula.[2] Thus, it is significant to understand the key issues discussed and the joint statement of principles signed by the six nations at the fourth round of the six-party talks.

The First Session of the Fourth Round of the Six-Party Talks

The first session of the fourth round of the six-party nuclear talks, held in Beijing on July 26 July-August 7, 2005, went into recess after the six nations engaged in intensive negotiations. The U.S. and the DPRK actively engaged in direct bilateral negotiations for thirteen days, but could not reach an agreement on the key issue of Pyongyang's right to peaceful nuclear activities.[3] This issue became a stumbling block to the consensus on a joint statement of principles by the six participants, involving the U.S., China, Russia, Japan, the DPRK, and the Republic of Korea (ROK or South Korea).

DPRK Position: Right to Peaceful Use of Nuclear Energy

DPRK Vice Foreign Minister Kim Kye-gwan demanded that the DPRK be allowed to retain its right to pursue peaceful nuclear activities. The DPRK expressed resentment towards the U.S.'s opposition to its peaceful use of nuclear power. Despite concerted diplomatic pressure, the DPRK consistently insisted on its right to a nuclear program for peaceful purposes. Pyongyang demanded the right to peaceful use of nuclear energy under the conditions that it would abandon all nuclear weapons programs, including the existing nuclear reprocessing and enriched uranium facilities in the future. Furthermore, North Korea wanted to restore the suspended the Korean Peninsula Energy Development Organization (KEDO) light-water nuclear reactor project. Both Seoul and Washington opposed this demand.

The DPRK's demands created problems for the ROK government. Seoul's offer to supply 2 million kilowatts of electricity to Pyongyang was based on the premise that the KEDO project be scrapped and Seoul's financial contributions to the KEDO would be redirected to the electricity supply project.

U.S. Position: No Nuclear Energy

The U.S. argued North Korea should not even be allowed to maintain nuclear reactors for civilian use because it turned a research facility at Yongbyon into a production center for weapons-grade plutonium after the collapse of the 1994 Geneva Agreed Framework. Light-water nuclear reactors (LWRs) are considered less likely to

2 For the Korean peninsula peace regime building initiative, see Tae-Hwan Kwak, "In Search of the Korean Peninsula Peace Regime Building," *Pacific Focus*, vol. 20, no. 2 (Fall 2005), pp. 147-92.

3 For details, see Tae-Hwan Kwak, "North Korea's Right to Peaceful Nuclear Activities," *Vantage Point*, vol. 28, no. 9 (September 2005), pp. 15-19.

produce weapons grade plutonium than graphite–type nuclear reactors do, but the U.S. still has nuclear proliferation concerns with such installations. Furthermore, President Bush supported South Korea's offer of electricity to the North as another reason why Pyongyang doesn't need any kind of nuclear program, even for power generation.

ROK Position: Conditional Support

Unification Minister Chung Dong-young said in an interview on August 11, 2005, "North Korea has a general right to peaceful use of nuclear energy for agricultural, medical and power-generating purposes." He also said, "In this, our position differs from that of Washington."[4] Chung's initial statement did not attach conditional support, but later President Roh included provisional support for the DPRK's right to use nuclear energy peacefully. ROK Foreign Minister Ban Ki-moon also said that the North should be allowed to explore peaceful uses of nuclear energy if it rejoins the Nuclear Non-Proliferation Treaty (NPT) and allows inspections by the International Atomic Energy Agency (IAEA). Ban reiterated that there was no major difference in principle between the U.S. and South Korean views on the DPRK's possible future use of nuclear energy.

Unification Minister Chung made it clear that the ROK would not supply two million kilowatts of electricity to North Korea and concurrently fund the KEDO project to construct two LWRs. He said the ROK electricity aid plan for North Korea was based on the premise that the DPRK would abandon its nuclear weapons programs and the KEDO project.[5] His remarks were intended to make it clear that the ROK would not endorse the DPRK's demand for restoring the suspended KEDO project if Pyongyang would accept Seoul's offer of electricity. Chung reaffirmed the ROK's conditional support for the DPRK's right to peaceful use of nuclear energy.

No compromise over this issue would mean the failure of the fourth round of the six-party talks. Even if the second session of the fourth round of the six-party talks resumed, it would be difficult to bridge the wide gap between Pyongyang and Washington.

The Second Session of the Fourth Round of Six-Party Talks

The second session was held in Beijing on September 13-19, 2005, when the six nations finally signed a joint agreement of principles for designing a detailed roadmap for achieving denuclearization of the Korean peninsula. The North Korean position remained unchanged. The key issue at the fourth round of six-party talks was North Korea's demand for a LWR to produce electricity as part of any deal to give up its nuclear weapons program. In the meantime, the Bush administration remained adamant that the DPRK must abandon its nuclear weapons programs and

4 *Yonhap News,* August 11, 2005.

5 Ibid.

forgo nuclear energy production. The U.S. refused to consider a light-water nuclear reactor for North Korea.

The ROK proposed supplying electricity to North Korea as part of the nuclear disarmament agreement, which chief U.S. negotiator Christopher Hill suggested, was a more practical way to meet North Korea's energy needs if electricity was the issue.[6] North Korea again demanded on September 14, 2005, that the U.S. and other nations give it money to build a new light-water nuclear reactor before it would end its nuclear weapons programs. Hill said, "Neither the United States nor any other participants is prepared to fund a light-water reactor."[7] A light-water reactor would cost 2-3 billion dollars and take about a decade to build. The five participants in the talk agreed that North Korea's condition that it would receive a LWR before ending its arms programs was unacceptable. However, the U.S. signaled a softening of that line, saying that they would be willing to leave aside the issue of civilian uses of nuclear technology for now, to clear the way for a general agreement on ending North Korea's nuclear weapons programs.[8]

Key Points of the September 19 (2005) Agreement

The joint statement was signed on the seventh day, of the second session, of the fourth round of the six-party talks. Although the accord included only general terms of principles, it marked the first specific agreement among the six parties since the six-party talks began in August 2003. It was designed to serve as the basis for further talks on the timing of North Korea's dismantlement of its nuclear weapons programs, the corresponding provisions of economic aid and diplomatic relations, and other incentives for the DPRK.[9] (See Appendix 2.1)

The joint statement was a diplomatic victory for China as host and mediator. The agreement was based on a compromise proposed by China to bridge differences between Washington and Pyongyang over the issue of a LWR. The compromise was achieved: the DPRK would be accorded the right to peaceful nuclear energy in principle, but only after dismantling its nuclear weapons programs and rejoining the U.N nuclear inspection regime and the NPT.

The Chinese compromise proposal was introduced after it became apparent that North Korea would not accept an earlier draft agreement with no mention of its demand for LWR as part of any accord on abandoning its nuclear weapons programs. The agreement said that "The DPRK stated that it has the right to peaceful uses of nuclear energy. The other parties expressed their respect and agreed to discuss at an

---

6 Edward Cody, "N. Korea Holds to Its Demands at Nuclear Talks," *Washington Post*, September 15, 2005, A28; Joseph Kahn, "North Koreans Insist on Demand for New Reactor in Nuclear Talks," *New York Times,* September 16, 2005.

7 Cody, *op. cit.*

8 Joseph Kahn, "North Korea Sets New Demand for Ending Arms Program: Money to Buy a Civilian Reactor," *New York Times*, September 15, 2005.

9 For details, see "Text of Joint Statement From Nuclear Talks," *New York Times*, September 19, 2005.

appropriate time the subject of the provision of a light-water reactor to the DPRK."[10] Those terms of the agreement represented concessions by the U.S. and the DPRK.

The DPRK Concessions

The agreement stated that "The DPRK committed abandoning all nuclear weapons and existing nuclear programs and returning at an early date to the treaty on the non-proliferation of nuclear weapons and to IAEA safeguards."[11] However, the agreement was vague at best, and did not specify anything about when or under what conditions the DPRK would dismantle all of its nuclear programs, re-enter the NPT, and allow IAEA inspections.

There is no mention about the HEU issue, which sparked the second nuclear crisis. The agreement does not explicitly address the issue of North Korea's uranium program. North Korea still denies having one, despite growing evidence that it at least tried to develop bomb fuel based on the HEU program with Pakistan's assistance.[12] However, the enriched uranium program was covered by the pledge to dismantle all nuclear weapons and existing nuclear programs and by a separate reference to a 1992 inter-Korean joint declaration on the Korean peninsula denuclearization, which prohibited uranium enrichment. But the accord did not require North Korea to confess the existence of the HEU program, meaning that unless the North admits to the program in a declaration of all nuclear facilities, international inspectors would have to work to uncover the uranium program in an adversarial way down the road. HEU will remain a key issue at the future six-party talks. Moreover, the joint statement has no mention of a verification process. Regarding the timing of the provision of LWR to North Korea, Secretary of State Rice argued that the wording of the agreement implies that North Korea would disarm first. She stated several times that the discussion on LWR would not even begin until North Korea dismantled its weapons programs.[13]

U.S. Concessions

After four years of bitter arguments over whether to negotiate with the DPRK or to try to engineer its collapse, President Bush finally approved the bare minimal agreement—an agreement, in principle, that the DPRK would abandon its nuclear weapons programs. Part of the reason Bush signed the agreement was that he had to close the breach among his negotiating partners: China, Russia, Japan and the ROK. Hard-liners in the Bush administration blocked negotiations for the first few years

---

10 Ibid.

11 Ibid.

12 For a review of the HEU program , see Bruce E. Bechtol, Jr., "Assessing the Present and Charting the Future of US-DPRK Relations: The Political Diplomatic Dimension of the Nuclear Confrontation," in Tae-Hwan Kwak and Seung-Ho Joo, eds., *The United States and the Korean Peninsula in the 21st Century* (Aldershot, England: Ashgate, 2006), pp. 119-23.

13 David E. Sanger, "Yes, Parallel Tracks to North, but Parallel Tracks Don't Meet," *New York Times*, September 20, 2005.

of Bush's presidency, but in the end, both President Bush and Secretary Rice were persuaded that if a confrontation ever occurred with North Korea, they had to show that they had made every effort to find a diplomatic solution. The Bush administration finally dropped its opposition to the DPRK receiving a LWR in the future, showing a softening of its hard-line position. President Bush cautiously welcomed the agreement, but warned, "We expect a verifiable process."[14] So verification will be a major issue at the future six party talks.

Chief U.S. negotiator Hill said that the administration didn't want to see any mention of providing North Korea with LWR in the joint statement. But the Chinese included it. The United States also balked at the use of the vague term "appropriate" to describe the timing. South Korea, Russia and China were happy to accept that language, because it left open the question of when the DPRK would receive the nuclear reactor. To break the impasse, Secretary Rice suggested that each country would issue separate statements describing their understanding of the deal with a specificity that is not in the agreement itself. The ROK and Japan went along with the idea, though Seoul complained that it would "sour the atmosphere." Russia and China issued vague statements that were unclear about the sequence of events.[15]

Overall, both President Bush and Chairman Kim Jong-Il were benefited by finally approving the joint agreement. The DPRK took a major step toward securing international acceptance. It will allow Chairman Kim to hang on to power for the foreseeable future, will gradually open North Korea to foreign investment, and avoid its sudden collapse. One long-term incentive in the joint statement was the call for Washington and Tokyo to take steps to normalize relations with Pyongyang if the DPRK dismantles its weapons programs. Normalization of diplomatic relations between North Korea and Japan could mean billions of dollars worth of economic assistance, from Japan alone, in belated World War II-era reparations. For President Bush, the agreement was welcome at a time when the war in Iraq lost considerable support at home and negotiations with Iran over its nuclear programs went astray.

Several controversial issues needed to be resolved at the future six-party talks. First, the U.S. urged the DPRK to issue a declaration of its nuclear weapons, materials, and facilities so that proper verification procedures could be devised. This request could be controversial because the U.S. wanted Pyongyang to include its HEU program in the declaration. Second, the U.S. and the DPRK will likely disagree with the proper sequencing for issues in the joint statement. Washington wanted Pyongyang to fulfill its denuclearization commitments before providing any benefits. By contrast, the DPRK maintained that all issues should be resolved on the basis of simultaneous actions. Third, the six nations needed to agree on verification procedures and the timing of LWR provision.

---

14 Ibid.

15 Joseph Kahn and David E. Sanger, "U.S.-Korean Deal on Arms Leaves Key Points Open," *New York Times,* September 20, 2005; Glenn Kessler and Edward Cody, "N. Korea, U.S. Gave Ground to Make Deal," *Washington Post*, September 20, 2005, p. A01.

The First Session of the Fifth Round of the Six-Party Talks

The first session of the fifth round of the six party talks held on November 9-11, 2005 to implement the 9.19 joint agreement ended without any progress. During the three days of discussions in Beijing, the parties presented their positions on implementation plans at the general session. A three-point plan to resolve the North Korean nuclear dispute was put forward on the opening day, breaking down the three categories of: (1) the dismantling of North Korea's nuclear program, (2) the economic cooperation and energy aid for North Korea, and (3) the normalization of relations between North Korea, the U.S., and Japan. The U.S. and Japan issued similar road maps, proposing the talks should be broken down into working-level groups to discuss the three categories. It was unclear if North Korea would accept the proposals.

Chief ROK negotiator Song Min-soon told reporters in Beijing that North Korea did not single out the issues one-by-one but came up with a rudimentary opinion about the issue of a light-water nuclear reactor and the process of dismantling the nuclear program.

During his trip to Brazil, President George W. Bush called North Korean leader Kim Jong-Il a "tyrant" again. North Korea's human rights issue, which was omitted in the joint statement, may become the hot topic. Washington's hardliners have been openly disapproving of U.S. chief negotiator Chris Hill's flexible approach to the North, which led to the fallout of his initial plan to visit Pyongyang before the fifth round of the six-party talks.

The DPRK will face many difficulties with diplomatic negotiations in the future. Human rights issues may well become one of the largest stumbling blocks for prospective negotiations. Along these lines, Japan has called for North Korea to amend the kidnapping case prior to normalizing diplomatic relations between the two states. Japan claimed that North Korea kidnapped 15 Japanese to train spies during the 1980s, while the DPRK insisted that it kidnapped 13 Japanese, of which eight died. North Korea sent five of them back to Japan. Tokyo demanded that Pyongyang provide more specific data of the eight deceased, claiming that the remains sent by North Korea could be bogus.

The U.S. Financial Sanctions: Major Obstacle to Resuming the Second Session of the Fifth Round of Six-Party Talks

The six-party nuclear talks came to a deadlock since the DPRK demanded that the U.S. lift its financial sanctions at the first session of the fifth round of the talks in November 2005. A brief analysis of: (1) U.S. financial measures influencing the negative impact on the six-party talks, (2) both U.S. and North Korean positions, and (3) a possible resolution of this issue will be presented.

Banco Delta Asia as a "Pawn" for Pyongyang

The U.S. Treasury Department in September 2005 designated Banco Delta Asia (BDA) in Macau as a financial institution of "primary money-laundering concern" under Section 311 of the Patriot Act. The U.S. claimed that BDA was a pawn for the North Korean government to engage in illicit financial activities through Macau. BDA allegedly provided financial services for more than 20 years to the DPRK and facilitated many of its criminal activities, including counterfeit U.S. currency. The U.S. launched a large-scale operation aimed at cracking down on North Korea's criminal activities and claimed that nearly all fake $100 bills, known as "supernotes," were believed to be produced by North Korea. The U.S. alleged that senior officials at BDA handled large cash deposits, including counterfeit US currency supplied by North Korean officials, and agreed to put the fake currency into circulation. The DPRK has allegedly produced more than $45 million in high-quality fake $100 bills since 1989.[16]

In October 2005, the U.S. blacklisted eight other North Korean firms for allegedly participating in the spread of weapons of mass destruction. In mid-December, the U.S. issued a formal advisory concerning North Korea's illegal activities and cautioned U.S. financial institutions to take steps to guard against the abuses of their financial services by North Korea. President Bush vowed to protect the U.S. currency and continue a crackdown on North Korea for fake U.S. dollar bills on January 26, 2005. He said, "There is no compromise."[17] Thus, the U.S.'s position on this issue was firm and uncompromising with North Korea.

BDA in Macau froze $24 million deposited in some fifty North Korean accounts. Other financial institutions also reportedly curtailed their dealings both with the bank and North Korea. Korea Exchange Bank (KEB), controlled by U.S. equity fund Lone Star, terminated all transactions, including remittance and foreign exchange dealings with BDA and KEB was the first South Korean financial institution to join U.S.-imposed sanctions against BDA. The two Japanese banks, the Bank of Tokyo-

---

16 For details, see the first statement issued by the US Treasury Department online at <http://www.ustreas.gov/press/releases/js2720.htm>; For see money laundering online at <http://usinfo.state.gov/ei/economic_issues/terrorist_financing/money_laundering.html>; For U.S. scrutiny for possible connections to North Korea's illicit fund-raising network, see Glenn R. Simpson, Gordon Fairclough, and Jay Solomon, "U.S. Probes Banks' North Korea Ties," *The Wall Street Journal*, September 9, 2005; For details, see Bill Gertz, "U.S. accuses North Korea of $100 bill counterfeiting," *Washington Times*, October 12, 2005; The document was posted on the U.S. Department of Justice website at <http://www.usdoj.gov> in October 2005; Bill Gertz, "N.Korea charged in counterfeiting of U.S. currency," *Washington Times*, December 2, 2005.

17 "Bush Vows Crackdown on N.Korea on Fake Dollars," *Reuters*, January 26, 2006; Carol Giacomo, "US rejects N.Korea demand to end finance crackdown," *Reuters*, January 4, 2006. online at <http://today.reuters.com/news/newsArticleSearch.aspx?storyID=19779+04-Jan-2006+RTRS&srch=us+crackdown>

Mitsubishi UFJ, the world's largest bank, and Mizuho Bank, Ltd., also terminated all transactions with BDA.[18]

Changing North Korean Position after Kim's Secret Visit to China

The DPRK vehemently denied U.S. allegations of counterfeiting and money laundering. The DPRK repeated that it would not return to the six-party talks unless the U.S. lifted its financial sanctions against it. The DPRK demanded an end to the U.S.'s crackdown on its financial activities before it would return to the six-party talks. Chairman Kim Jong-Il reaffirmed his commitment to the six-party talks and, again, to denuclearization during his secret visit to China on January 10-18, 2006. Kim and Chinese President Hu Jintao agreed to maintain the "stand of seeking a negotiated peaceful solution" to the nuclear issue by pledging to work together to "overcome the present difficulties" at the six-party talks.[19]

The long-stalled six-party nuclear talks will not serve the interests of the parties concerned, particularly the U.S. and the DPRK. It would impede North Korea's effort to revive its moribund economy with foreign investment. And for Washington, it would obstruct the U.S.'s effort to stop the spread of North Korean nuclear weapons. According to Kyodo News on February 11, 2006, Chairman Kim Jong-Il said his government could collapse if the United States continued to impose financial sanctions against North Korea. Kim made the remarks when Chinese President Hu Jintao asked him in Beijing in January 2006 to drop the lifting of the U.S. sanctions as a condition for North Korea to return to the stalled six-party talks.[20] It was clear that this indicated Chairman Kim's fear of U.S. financial sanctions against North Korea.

Different Approaches to the BDA Issue

DPRK position The DPRK was willing to make a compromise with the United States. Chief U.S. delegate Christopher Hill met with the Chinese and North Korean delegates in Beijing on January 18, 2006. According to Hill, the DPRK "indicated they would be prepared to subscribe to international norms with respect to money laundering and would want to cooperate internationally on these issues."[21] A DPRK Foreign Ministry spokesman, on February 9, 2006, insisted that his government did not sanction crimes enumerated by the U.S. and even suggested that individuals responsible for them would be punished. Pyongyang again denied U.S. accusations of financial crimes, calling them fabrications aimed at defaming the North Korean

---

18 KEB joins sanctions against N.K. online at <http://www.koreaherald.co.kr/site/data/html_dir/2006/02/04/200602040008>.

19 For details, see Philip P. Pan, "In China, Kim Vows Commitment to Talks," Washington Post, January 19, 2006.

20 "N. Korea could collapse if U.S. sanctions continue: sources," February 11, 2006, online at <http://home.kyodo.co.jp/modules/fstStory/index.php?storyid=229607>.

21 Carol Giacomo, "N.Korea hints at curbing money laundering," Reuters, January 25, 2006.

regime. He stressed, "North Korea opposes all sorts of illegal acts in the financial field ... We will, as ever, actively join the international actions against money laundering."[22] This important message indicated that the DPRK was ready to make a compromise with the U.S. on the issue of counterfeiting and money laundering.

Li Gun, Director of American affairs in DPRK Foreign Ministry, made four requests at the March 7, 2006 meeting in New York. He demanded that the United States lift its financial sanctions, establish a joint U.S.-DPRK team to examine the counterfeiting concerns, give North Korean access to the U.S. banking system, and provide North Korea with technical assistance on identifying counterfeit bills.[23] He was reported as saying that Pyongyang would not link the U.S. financial measures with the six-party talks, according to Kyodo News.[24] However, he said that the lifting of the U.S. sanctions against BDA was "the least condition" necessary for Pyongyang's return to the six-party nuclear talks, raising hopes that the DPRK would return to the six party talks if the U.S. would be flexible on the BDA issue.

U.S. position The U.S. Department of State on February 10, 2006 dismissed North Korea's denunciation of illicit financial activities as "rhetoric" and demanded Pyongyang should show action by ceasing illegal activities immediately. State Department spokesman Sean McCormack said at a daily briefing, "Well, it's a fine rhetorical commitment, but what we would call upon is the North Korean government to cease all such activities," and, "It's very simple. It is within the power of the North Korean government to do so."[25] He said on February 15, 2006, that the DPRK must show some "convincing evidence" that it had stopped counterfeiting US currency to satisfy U.S. concerns. Specifically, he stated that Pyongyang must "provide evidence that the equipment and plates for the so-called super notes had been destroyed so that concerns about further ability [to print more notes] will be reduced."[26] Washington waited to see if Pyongyang would take concrete action to prove it would stop its illicit financial activities.

At the March 7, 2006 New York meeting, Assistant U.S. Treasury Secretary Daniel Glaser gave a briefing on U.S. actions on Pyongyang's illicit financial activities for Li Gun. The U.S. rejected a DPRK's proposal for establishing a separate negotiation framework bilaterally with Washington to discuss its punitive measures against BDA and the allegations against North Korea.[27] The U.S. again urged North Korea to return to the six-party talks. It was reported that the U.S. urged North Korea to

22 KCNA, February 9, 2006; "North Korea vows to join international anti-money laundering drive," February 9, 2006, online at <http://mdn.mainichimsn.co.jp/international/asia/news/p20060209p2g00m0in029000c.html.

23 Glenn Kessler, "N. Korea Sets Terms for Return to Nuclear Talks," *Washington Post*, March 9, 2006; "N.K. proposes separate talks on sanctions," *Korea Herald*, March 9, 2006; Lee Dong-min, "U.S. will look at Pyongyang's proposal on BDA talks," *Yonhap News*, March 9, 2006.

24 *Kyoto News*, March 9, 2006.

25 See US response to the DPRK's statement of February 9, 2006 online at <http://english.yna.co.kr/Engnews/20060211/630000000020060211053551E6.html>.

26 "U.S. Demands Proof of End to Fake Bills," *Joongang Ilbo*, February 16, 2006.

27 *Yonhap News*, March 11, 2006.

join the Asia-Pacific Group on Money Laundering (APG), a regional anti-money laundering organization, to demonstrate its commitment to prevent illicit financial activities.[28] The U.S. clearly stated its position: the issue of North Korea's money laundering and counterfeiting activities was non-negotiable.

The Bush administration believed that U.S. financial sanctions against North Korea had proved to be far more effective in causing a ripple effect around world than anyone had imagined. Banks around the world have limited their dealings with North Korea. The White House gave the Treasury and Justice Departments "full authority to take additional legal and financial actions against North Korea."[29]

U.S. effective action against North Korea once again energized hard-liners in the Bush administration. The U.S. appeared to have adopted a new policy, "squeeze them [North Koreans], but keep the negotiations going."[30] The U.S. new policy was likely to incite North Korea to take similar actions, and moderates in the State Department feared that it would not persuade North Korea to disarm, while also alienating Beijing and Seoul.

ROK National Intelligence Service (NIS) Director Kim Seung-gyu said in early February 2006 that it had no conclusive evidence that the DPRK had been counterfeiting U.S. dollars since 1998.[31] A widening gap in the U.S.'s and the ROK's strategies for dealing with the DPRK has produced a disagreement between Washington and Seoul. While Washington favored a harder line, Seoul desired a "pragmatic" approach to North Korea. The ROK seemed to fear that pushing the counterfeiting issue may derail the six-party process.

China said that U.S. financial sanctions against the DPRK had a negative impact on the six-party talks, stressing the need for active contacts between all parties concerned in order to promote the six-party process. China investigated the North Korean counterfeiting and money laundering issue. But it has not made any public statement regarding the findings of the investigation. It appeared that the U.S. believed that China had accumulated clear evidence about North Korea's counterfeiting. Nevertheless, China was reluctant to provide it to the U.S. because it did not want to push North Korea to a dangerous level, risking the North Korea's collapse.

A New Approach to North Korea

As discussed above, the United States took a rather flexible approach to the North Korean nuclear issue at the second session of the fourth round of six-party talks in September 2005. Thus, as a result, the 9.19 joint statement was signed by the

28 "U.S. asks NK to join int'l group on money laundering," *Yonhap News*, March 12, 2006.

29 Joel Brinkley, "U.S. Squeezes North Korea's Money Flow," *New York Times*, March 10, 2006.

30 Ibid.

31 Kwang-tae Kim, "Agency: North Korea Not Counterfeiting," *Associated Press*, February2, 2006, online at <ttp://www.washingtonpost.com/wp-dyn/content/article/2006/02/02/AR2006020200244.html>.

six nations. After the 9.19 joint statement was signed, the United States put more pressure on North Korea along with new issues of alleged counterfeit, drug trafficking, and human rights. Nevertheless, the U.S.'s pressure on North Korea was counter-productive and the DPRK's reaction was even hostile toward the U.S.

The Bush administration began realizing that its hard-line policy, or "benign neglect," toward the DPRK has been ineffective. Therefore, some argue that it is desirable for the U.S. to change its hard-line policy toward North Korea to implement the 9.19 joint agreement.[32] Thus, U.S. policy change toward the DPRK was signaled at the March 7, 2006 meeting in New York. The U.S. proposed bilateral talks with North Korea in the framework of the six-party talks. In addition, leaders in the U.S. Congress urged President Bush to offer direct U.S.-DPRK bilateral talks, a peace treaty, and normalization of diplomatic relations between the two countries.

In May 2006, President Bush's top advisers recommended a new approach to dealing with North Korea that would include beginning negotiations on a peace treaty. The new approach was hotly debated among different factions within the Bush administration, and reportedly President Bush will not approve it unless North Korea returns to the long-stalled six-party talks. For many years, the Bush administration maintained that it would not improve its relations with North Korea until it completely and verifiably dismantled its nuclear program. That stance later softened, and the administration said some benefits to North Korea could begin to flow as significant dismantlement took place. Now, if Bush allows talks about a peace treaty to take place on a parallel track with the six-party talks, it will signal another major policy change vis-à-vis Pyongyang.[33]

The U.S. Congress urged the Bush administration to review its "benign neglect" policy toward the DPRK. The U.S. Senate on June 22, 2006, voted for an amendment to the defense authorization bill, demanding the appointment of a new senior envoy to review the Bush policy toward Pyongyang. The amendment requires that within 60 days after the defense bill becomes law, the president shall appoint a senior presidential envoy as coordinator of U.S. policy on North Korea.[34] It was approved by the U.S. House of Representatives to become law.

The U.S. policy change was influenced, in part, by growing concerns about Iran's nuclear program and the long-stalled six-party talks. Secretary of State Condoleeza Rice instructed her top aide Philip D. Zelikow to draft two papers describing the new approach to North Korea. Those papers were reportedly discussed by President

---

32 For a critique of the Bush administration's policy toward North Korea, see Leon V. Sigal, "An Instinct for the Capillaries," paper presented Seoul-Washington Forum, co-hosted by the Sejong Institute and the Center for Northeast Asian Policy, the Brookings Institution, Washington DC, May 1, 2006 online at <http://www.brook.edu/comm/events/20060501_sigal.pdf> (November 3, 2006); Kenneth Quinones, "US-Japan Pressure and Tokyo-Seoul Tensions Help Pyongyang," *Mainichi Shimbun,* June 15, 2006; Michael E. O'Hanlon and Charles L. Prichard, "A New Approach to North Korea," *The Baltimore Sun*, April 12, 2005, p. 4.

33 David E. Sanger, "U.S. Said to Weigh a New Approach on North Korea," *New York Times,* May 18, 2006; "Bush may offer new carrot to end Korean nuclear crisis," AFP, May 18, 2006.

34 "U.S. Senate calls for new N. Korea envoy, review," *Reuters*, June 23, 2006.

Bush and Vice President Cheney, who was widely described by current and former officials as leading the drive in Bush's first term to make sure the DPRK received no concessions from the U.S. until all of its weapons and weapons sites were taken apart.[35]

North Korea's Missile and Nuclear Tests and Implications for Northeast Asian Security

North Korea's Ballistic Missile Tests

The U.S., Japan, South Korea, China, and Russia strongly urged the DPRK not to test-fire a long-range ballistic missile. Some American leaders favored a proposal for a preemptive strike on the North Korean missile sites if North Korea tested missiles, while others and the Bush administration opposed it.[36] But on July 5, 2006, the DPRK shocked the world by launching seven missiles, including an unsuccessful launch of its Taepodong-2 long-range missile. The short and medium-range missiles were successfully launched. The DPRK had reportedly informed China of its missile launch plan prior to its missile tests, but China denied the report. The DPRK held that the launchings of the seven missiles had been "routine military exercises." It declared that it would continue to test-fire missiles. Thus, the DPRK Foreign Ministry justified the missile launches which were "part of the routine military exercises staged by the KPA to increase the nation's military capacity for self-defense."[37]

The DPRK wanted to draw the attention of the U.S., which was being increasingly distracted by an Iranian nuclear weapons program, and wanted to send a strong message to Washington that it would not continue to tolerate U.S. financial sanctions against Pyongyang. North Korea had been quite frustrated by the U.S. rejection of its demand for direct, bilateral talks. Thus, North Korea appeared to use ballistic missile tests as a bargaining chip to induce the U.S. to come to direct talks.

The North Korea's test launch of a long-range ballistic missile significantly had a negative impact on peace and stability in the Korean peninsula and Northeast Asia, the missile defense system, the long-stalled six party talks, Japan-DPRK relations, and U.S.-ROK relations. However, it strengthened the U.S.-Japan alliance.[38] The U.N. Security Council moved toward a vote on a resolution sanctioning North Korea for its missile launch, despite dissent from China and Russia. Both nations

---

35 David E. Sanger, "U.S. Said to Weigh a New Approach on North Korea," *New York Times,* May 18, 2006.

36 For details, see Ashton B. Carter and William J. Perry, "If Necessary, Strike and Destroy," *Washington Post*, June 22, 2006, A29; David E. Sanger, "Don't Shoot. We're Not Ready, " *New York Times*, June 25, 2006; Glenn Kessler, "U.S. Rejects Suggestion to Strike N. Korea Before It Fires Missile," *Washington Post*, June 23, 2006, p. A21; Charles L. "Jack Pritchard, No, Don't Blow It Up," *Washington Post,* June 23, 2006, p. A25.

37 "DPRK Foreign Ministry Spokesman on Its Missile Launches," KCNA, July 6, 2006, online at <http://www.kcna.co.jp/index-e.htm (November 21, 2006)>.

38 Joseph Coleman, "Missile Threat Strengthens U.S.-Japan Ties," *Associate Press*, June 23, 2006.

resisted US pressure to take a get-tough approach to North Korea because it could fuel instability and jeopardize efforts to restart six-party talks. However, neither country has threatened to use its veto power to block U.N. sanctions. Finally, the U.N. Security Council voted for its resolution 1695 sanctioning North Korea.[39]

Assistant Secretary of State Christopher Hill made criss-crossing visits to Beijing, Seoul, Tokyo, and Moscow to gather support for coordinated pressure on Pyongyang. The U.S. pushed for a unified response to North Korean missile tests, but was only given lukewarm support from China and Russia. Seoul began to take some punitive measures against Pyongyang by rejecting the latter's proposal for military talks, made several days before the missile tests, and by holding off sending 500,000 tons of rice and 100,000 tons of fertilizer to North Korea. But inter-Korean ministerial meetings scheduled for July 11-14 were held as planned.

In the meantime, Han Song-ryol, deputy chief of DPRK's mission to the U.N., again said on July 7 that Pyongyang was willing to return to the six-party talks if the U.S. lifted punitive measures against the BDA and that the lifting of the measures would serve as proof of the U.S.'s will to resolve the current standoff through dialogue. He warned again that any U.N. sanctions against North Korea would be considered an "act of war." But the U.S. rejected Pyongyang's demand, arguing that the six-party talks and financial restrictions were separate issues and should not be linked.

North Korea's Nuclear Test and Its Negative Effects

The six-party process was virtually dead, primarily due to inflexibility and intransigence of both Pyongyang and Washington. After November 2005, the DPRK boycotted the six-party talks in protest against of U.S. sanctions against North Korea.

In defiance of stern international warnings by the U.S., China, Japan, Russia, the ROK, and the other members of the United Nations, the DPRK announced on October 9, 2006 had successfully conducted its first underground test of a nuclear weapon for its self-defense capability.[40] The nuclear test was secretly done only six days after the DPRK announced that it would detonate a nuclear device.[41]

Pyongyang has long proposed direct bilateral talks with the U.S. to discuss its nuclear and missile programs, but Washington rejected the proposal, arguing it would only have bilateral meetings with the DPRK within the framework of the six-

39 For details, see "Security Council Condemns Democratic People's Republic of Korea's Missile Launches, unanimously adopting resolution 1695 (2006)," online at <www.un.org/News/Press/docs/2006/sc8853.doc.htm> (December 20, 2006).

40 "DPRK Successfully Conducts Underground Nuclear Test," *KCNA*, October 9, 2006, online at < http://www.kcna.co.jp/index-e.htm> (October 14, 2006); David E. Sanger, "N. Korea Reports 1st Nuclear Arms Test," *New York Times*, October 9, 2006.

41 For clarifying DPRK Foreign Ministry's stand on new measure to bolster war deterrent, see "DPRK Foreign Ministry Statement," *KCNA,* October 3, 2006, online at <http://www.kcna.co.jp/index-e.htm>http://www.kcna.co.jp/index-e.htm> (October 14, 2006); David E. Sanger, "North Koreans Say They Plan a Nuclear Test," *New York Times*, October 4, 2006.

party talks. Chairman Kim Jong-Il might have calculated that a nuclear-armed North Korea could more easily induce the U.S. to direct bilateral negotiations.

The North Korea's nuclear test promptly gave rise to the U.N. Security Council resolution 1718, under Chapter 7, Article 41 of the U.N. Charter, aimed at punishing North Korea for its nuclear test by imposing economic sanctions on Pyongyang, thereby deeply isolating it from the international community.[42] The DPRK rejected the resolution, threatening "physical countermeasures" against any state that tried to enforce the U.N. sanctions.[43] The North Korean nuclear test failed to change the U.S.'s position on direct bilateral talks, and instead put more pressure on Pyongyang through the U.N. Security Council resolution. Furthermore, it posed a grave security threat to the Korean peninsula and Northeast Asia.

Pyongyang's Motives of Its Nuclear Test

Why, then, did North Korea want to go nuclear? What were its motives? Four factors explaining the nuclear test will be briefly discussed. First, the survival of the Kim Jong-Il regime is a top priority. The DPRK has repeatedly emphasized the buildup of its nuclear deterrent force against a US preemptive attack, while the U.S. has again and again stated that it has no intention to attack it. The North Korean fear of a U.S. attack is deeply rooted in its mentality.

Second, the North Korean nuclear test is viewed as a bargaining chip to strengthen its negotiation position with the U.S. Pyongyang conducted a nuclear test to put pressure on Washington to accept direct bilateral talks in order to find a way to lift financial sanctions against it. Using the brinkmanship tactics as in the past, Pyongyang attempted to force Washington to come to the negotiating table ahead of its mid-term elections in November 2006.[44]

Third, North Korea's economic motives also prompted its decision to go nuclear to earn hard currency by selling missiles, nuclear technology, and nuclear material to any potential buyers.

Fourth, the DPRK wanted to enhance its international prestige as a nuclear power, while getting revenge on the U.S. for its malign neglect, just as the 1998 Indian and Pakistani nuclear tests elevated their international prestige as nuclear powers. President Bush's 2002 "axis of evil" speech, the U.S. invasion of and subsequent quagmire in Iraq, the U.S. military doctrine of preemptive attack, and the

---

42 For details, see "Resolution 1718 (2006) Adopted by the Security Council at its 5551st meeting, on 14 October 2006," online at <http://daccessdds.un.org/doc/UNDOC/GEN/N06?572/07/PDF/N0657207.pdf?OpenElement (November 21, 2006)>; John O'Neil and Norimitsu Onishi, "U.S. Confirms Nuclear Claim by North Korea," *New York Times*, October 16, 2006; Colum Lynch and Glenn Kessler, "U.N. Votes To Impose Sanctions On N. Korea,Council Demands End To Nuclear Program," *Washington Post*, October 15, 2006, p. A01.

43 "DPRK Foreign Ministry Spokesman Totally Refutes UNSC 'Resolution,'" *KCNA*, October 17, 2006, online at <http://www.kcna.co.jp/index-e.htm. (10/23/2006)>.

44 Sohn Suk-joo, "N. Korea's nuclear test threat targets U.S. concessions," *Yonhap News*, October 3, 2006.

fear of a possible U.S. invasion may have contributed to Chairman Kim's decision to accelerate the nuclear development programs.

Debate over who should be blamed for North Korea's nuclear test is counterproductive: President Bush, Chairman Kim, ROK President Roh, PRC (People's Republic of China) President Hu, and virtually every participant in the six-party talks should be responsible for it.

Nuclear Test and Its Impact on Inter-Korean Relations

North Korea claimed a successful underground nuclear bomb test and joined the once exclusive club of nuclear powers as the ninth member. Pyongyang was a self-declared nuclear power in February 2005. The U.S., Japan, the ROK, and China asserted that a nuclear-armed North Korea would be neither tolerated nor recognized.[45] They insisted that Pyongyang give up its nuclear ambitions because a nuclear North Korea posed a grave threat to peace and security in Northeast Asia.

Inter-Korean relations have dropped to their lowest ebb since the inter-Korean summit in Pyongyang in June 2000, due to the North's nuclear test. Seoul has temporarily stopped its regular food and fertilizer aid to Pyongyang since the missile tests in July 2006, although it shipped a one-time aid package to help North Korean flood victims in the summer of 2006. The nuclear test spoiled inter-Korean economic cooperation projects such as the Gaesong industrial complex and the Mount Geumgang tourism resort. The high-level inter-Korean dialogue will be put on hold indefinitely, and there will be little chance that the inter-Korean family reunion project will resume for many months.

The nuclear test compelled the ROK to re-evaluate its engagement policy of national reconciliation and cooperation with North Korea.[46] The debate over the engagement policy led to the internal division in the South Korean society. It could be a hot campaign issue for the ROK's 2007 Presidential elections.

In the short-term, the nuclear test appeared to weaken the China-DPRK alliance, but it strengthened U.S.-China cooperative relations and the U.S.-ROK alliance system. The long-term impact on the Northeast Asian security will be significant.

Undermining of Nuclear Non-Proliferation Regime/Nuclear Arms Race

The North Korean nuclear test was a major challenge to the non-proliferation regime, and opened the Pandora box of a more dangerous nuclear age in the twenty-first century. It could spark a wave of nuclear proliferation in the world. A host of nations in the Middle East and Northeast Asian regions will be considering going nuclear, although U.S. assurance of extended deterrence to its allies remains firm. In the Northeast Asian region, the United States' extended deterrence has been effective. The long-standing U.S. commitment to the security of Japan and the ROK, based

45 Melinda Liu, "China's Reaction: Tightening the Screws, *Newsweek*, October 30, 2006.

46 "South Korea imposes sanctions on North Korea," *North Korea Times,* October 26, 2006.

on U.S. bilateral defense treaties with them, has been firm and has reduced their incentives to go nuclear at present. Thus, Japan and the ROK will continue to rely on the U.S.'s nuclear umbrella in the short-term, despite the strategic imbalance resulting from the North Korean nuclear test.

Japan and the ROK feel secure under the U.S. extended nuclear deterrence for now. But in the long run, they may feel insecure if North Korea proves capable of putting a nuclear warhead on a missile, if it could blackmail or nuke Japan, and if it could strike the U.S mainland with a nuclear warhead. Japanese and the ROK's leaders would ask the U.S. if it is willing to risk a possible nuclear war with North Korea. Japan and the ROK will go nuclear if the U.S.'s credibility is in question. Thus no one can be sure that they would not eventually go nuclear in the long-term. The same is true of Taiwan.

The North Korean nuclear test will prompt Iran to acquire the capability to produce its own nuclear weapons. Iran insists it is interested only in a civilian nuclear program for energy purposes. The US, the European Union, and the IAEA are increasingly skeptical about those claims, but thus far have been powerless to do much about it. It is widely assumed that Iran could become the next nuclear power if it proceeds undeterred with its clandestine nuclear program. The North Korean test suggested that nobody would stop Iran from going nuclear. If Iran goes nuclear militarily, others will follow suit. Turkey would think seriously about going nuclear if Iran, Israel, and Turkey were all nuclear powers. The Arab states, including Saudi Arabia, would feel they have no choice but to follow suit.

Japan stated it would not go nuclear for now. However, the North Korean nuclear test certainly prompted Japan to develop a MD system in cooperation with the U.S. In turn, it would push Beijing to spend more money on developing advanced nuclear weapons. Thus, a possible new nuclear arms race among Japan, China and other countries in Northeast Asia would soon start.

The worst-case scenario is that North Korea or Iran will sell nuclear weapons to a terrorist group. If Kim Jong-Il were to sell a nuclear weapon to Bin Laden and that weapon were used against the US or one of its allies, then the US would likely retaliate militarily against Pyongyang.

Effectiveness of U.N. Economic Sanctions

Many experts have argued that the UN economic sanctions against the DPRK may be counterproductive, particularly without full support from China, Russia, and the ROK. The effectiveness of the economic sanctions depends on member states' willingness to faithfully implement the U.N. resolution 1718. However, it should be noted that the pattern of North Korean behavior in response to outside pressure has shown that Pyongyang has not been yielding; instead it has been strengthening its self-defense capabilities by developing and testing missiles and nuclear devices.[47]

The U.S. has imposed sanctions against North Korea since the 1950s, but they have not prevented Pyongyang from acquiring long-range missiles and nuclear

---

47 Fareed Zakaria, "Let Them Eat Carrots," *Newsweek*, October 23, 2006; "Will Sanctions Stop North Korea?" *New York Times*, October 17, 2006.

weapons, nor have they led to the collapse of the regime. Will the U.N. imposed sanctions be effective and successful? Two factors should be considered in this regard. First, North Korea is already the most isolated country in the world. The majority of the North Korean people live below subsistence levels, and they are prepared to endure another "arduous march" ahead as they did in 1996-99.

Second, it depends on whether China and the ROK actively participate in the sanctions. China holds the key to effectiveness of the U.N. sanctions. North Korea's only substantial trade, including illicit trade, is with China. Much of the trade is on a concessionary basis, and if sanctions should include fertilizers, massive starvation is likely to recur in North Korea.

China will probably continue to make harsh remarks about North Korea by carrying out the U.N. Security Council resolution, but it does not want North Korea's collapse. In addition, China wants to maintain stability on the Korean peninsula and Northeast Asia for the 2008 World Olympics in Beijing and for its stable economic development. China has an economic leverage over North Korea, but it is reluctant to use it. If China cuts off its oil supply to North Korea, North Korea's economy will soon be paralyzed. China and the ROK were under intense pressure from the US to fully participate in the Proliferation Security Initiative (PSI), but they decided not to join it.

What should be done to induce North Korea to return to the six-party talks? Direct bilateral talks between the U.S. and the DPRK could still help jump start the six-party process. But the U.S. has rejected direct bilateral talks with Pyongyang. In September 2006, U.S. Ambassador to Seoul Alexander Vershbow indicated a slight shift in the U.S. position on bilateral talks. He said there was a possibility of a face-to-face meeting between the U.S. and DPRK negotiators if North Korea would also agree to return to the six-party talks.[48]

Direct bilateral talks between the U.S. and the DPRK are the key to the resolution of the current crisis. But the Bush administration's position is obstinate: The U.S. has already agreed to bilateral talks in the context of the six-party process, but has rejected direct bilateral talks.

The U.S. has urged the DPRK to return to the six-party talks without preconditions, just as the entire international community, including two UN Security Council resolutions, has urged it repeatedly to do. The DPRK's position is equally unyielding. Pyongyang wanted the U.S. to unfreeze $24 million in its accounts at the BDA as a precondition for its return to the six-party talks. But it has now softened its position on this issue. Pyongyang has suggested that it would return to the six-party talks in exchange for discussions on the BDA issue at direct bilateral talks. However, the U.S. has not changed its position and urged the DPRK to unconditionally return to the six-party talks to implement the 9.19 joint statement.

The DPRK will return to the six-party talks if the U.S. accepts Pyongyang's somewhat softened proposal for discussions on the frozen BDA funds at bilateral talks within and/or outside the framework of the six-party talks. Even if Pyongyang returns to the six-party talks, there will be long, rough, and frustrating negotiations

48 Lee Chi-dong, "Vershbow hints at Washington's flexibility on Pyongyang," *Yonhap News*, September 21, 2006.

over the 9.19 joint statement to denuclearize the Korean peninsula. The DPRK has repeatedly maintained that it would give up nuclear weapons in return for guarantees on its regime and economic aid. To achieve this goal, there should be a compromise between the U.S. and the DPRK. Both Washington and Pyongyang should demonstrate their political will to make a compromise in order to resolve the security issues posed by a nuclear North Korea.

The Second Session of the Fifth Round of the Six-Party Talks: Evaluation and Policy Recommendation

The second session of the fifth round of the six-party talks held in Beijing on December 18-22, 2006 went into recess without any breakthrough. The stalled six-party talks raised serious questions about its usefulness, according to U.S. major newspapers.[49]

The DPRK had boycotted the six-party talks since November 2005 in protest against a U.S. crackdown on Pyongyang's counterfeiting of U.S. dollars and money-laundering, which led to the freezing of $24 million in North Korean accounts deposited at the BDA in Macao. During the 13 month boycott, North Korea test-fired seven missiles in July 2006, and conducted its first nuclear test in October 2006, leading to two unanimously adopted U.N. resolutions against North Korea (1695 and 1718). The DPRK decided to return to the six-party talks, partly because of the U.S. agreement to discuss the BDA issue at bilateral U.S.-DPRK talks. Given North Korea's insistence that the United States first agree to lift financial sanctions against it before the nuclear issue could be discussed, expectations for any significant breakthrough were generally low. Thus, the second session appeared doomed from the start with conflicting approaches to the financial issue.

During five days of the talks in Beijing, Pyongyang refused to address its nuclear weapons but instead persistently demanded that U.S. financial restrictions be lifted. As a result, the six-party talks ended without any progress. But Chinese chief delegate Wu Dawei stated all six participants reaffirmed the 9.19 joint agreement and pledged to "reconvene at the earliest opportunity."

North Korea's Position

DPRK chief negotiator Kim Kye-gwan, on arrival in Beijing, demanded an end to the financial sanctions as a prerequisite to successful implementation of the 9.19 agreement. He said that the DPRK would not abandon its nuclear weapons until the United States gives up its "hostile" policy and drops financial sanctions. He also

---

49 For details of this round of the six-party talks, see Edward Cody, " Nuclear Talks With N. Korea End in Failure, Six-Party Process Thrown Into Doubt," *Washington Post*, December 23, 2006, p. A12; Glenn Kessler , "3 Crises Face U.S., With Risky Options Iran, Sudan, N. Korea Tactics Stall, But New Directions Hold Pitfalls," *Washington Post*, December 23, 2006, p. A14; Joseph Kahn, "Talks End on North Korea's Nuclear Weapons," *New York Times*, December 23, 2006; Mitchell Landsberg, "N. Korea nuclear talks end with no resolution," *Los Angeles Times*, December 23, 2006.

said, "The most important issue is for the U.S. to make a switchover in its policy" and "The problem will be resolved when the hostile policy is changed to a policy of co-existence. I do not yet know whether the U.S. is prepared to do that."[50] Thus, the DPRK has justified its possession of nuclear weapons because of the "hostile" policy of the United States, citing the financial restrictions, criticism of North Korea's human rights record, and U.S.-ROK joint military exercises. The North insisted on the lifting of the financial restrictions on North Korea to prove that the U.S. has changed its hostile stance.

In an opening statement in this session, Kim Kye-gwan laid out a set of conditions, including lifting U.N. sanctions and the BDA issue. But the North took a more pragmatic approach in subsequent talks. Sharp disagreements between the U.S. and the DPRK remained, especially over the financial issue. The DPRK refused to engage in negotiations on the nuclear issue until the BDA issue would be resolved. The DPRK insisted that the U.S. first lift its financial restrictions. In statements leading up to this session, North Korea repeated it would negotiate on its nuclear weapons program only after the lifting of restrictions imposed on several of some 50 North Korean accounts. In short, the DPRK was not interested in negotiating with the United States about the nuclear weapons programs, but instead North Korea focused upon the solution to the BDA issue.

U.S. Position

U.S. envoy Christopher Hill said the U.S. financial curbs issue would be dealt with in discussions carried out in parallel with the six-party talks. He said the six-party talks should focus on dismantling North Korea's nuclear programs, stressing that "It's very important that we not focus on those financial issues but rather on the central matter of denuclearizing the Korean peninsula."[51]

The United States insisted the nuclear and BDA financial issues would be resolved separately. Hill made a new U.S. proposal to North Korea, but it was not officially revealed in detail. The U.S. had reportedly outlined a process of the North Korea's nuclear dismantlement plan in which Pyongyang, at the initial stage, would first freeze its nuclear program and allow IAEA inspections, followed by the North's voluntary declaration of its nuclear programs, verification measures, and eventual dismantlement. In return, the U.S. offered security guarantees, a peace treaty, and normalization of relations, as well as removal of Pyongyang from Washington's list of states sponsoring terrorism, according to South Korean officials.[52]

The DPRK has yet to make any official response to the U.S. proposals. It reportedly said it would be willing to halt operation of its main nuclear reactor and

---

50 "Before Talks, North Korea Accuses U.S. of 'Hostile' Policy," *Reuters,* December 17, 2006.

51 Brian Rhoads, "U.S. presses for real progress in N.Korea nuclear talks," *Reuters*, December 16, 2006.

52 Kwang-Tae Kim, "SKorea Urges North to Mull U.S. Proposal," *Associate Press*, January 10, 2007.

allow international inspectors "under the right conditions."[53] The DPRK made it clear that it would only discuss a freeze on nuclear weapon production programs, and that it would not discuss giving up nuclear weapons it has already built.

The U.S. argued the financial issue had no direct relationship to the six-party nuclear talks. But North Korea's Trade Bank President Oh Kwang-chul and Deputy Assistant Secretary of Treasury Daniel Glazer had two-day working-level meetings in Beijing to discuss the BDA issue within the six-party framework. Those sessions were useful, but did not resolve the financial standoff. There have been signs that the U.S. might be willing to unlock part of the frozen $24 million funds. Glaser said a resolution of the financial issue would require a "long term process."

After preliminary contacts in Beijing, the U.S. and the DPRK decided to continue talks in New York in January 2007. But North Korean negotiators said the BDA issue must be resolved before they can begin official talks on implementing the September 2005 agreement. The BDA issue was a major interest for North Korea. North Korea's demand emboldened by its nuclear status could no longer be ignored by the U.S. The U.S. and other participants believed they could go forward while the BDA issue was resolved on a parallel track. Hill said that the North Korean delegation "did not have the instruction" needed to go forward."[54]

Usefulness of the Six-Party Talks

The usefulness of continuing the six-party talks was questioned, given North Korea's stance on the BDA issue, suggesting that the six-party talks format could be scrapped after more than three years of inconclusive results. Some U.S. officials have concluded that the slow negotiations at the six-party talks have simply let North Korea stall for time, allowing it to build up its nuclear weapons and ultimately conduct a nuclear test. Hill said, "We are disappointed that we were unable to reach any agreement," and "Diplomacy is not an easy task, but like many things in life, you have to look at the alternatives." And he also said, "We still believe that diplomacy is the best way to solve this, and we believe in particular that the six-party process is the best way to solve this."[55]

Many wondered whether Chairman Kim Jong-Il made his firm determination to abandon nuclear weapons. North Korea will not easily give up its nuclear arsenal, unless the U.S. and the international community will meet Pyongyang's demands. DPRK envoy Kim Kye-gwan said the U.S. is using a carrot-and-stick approach to his government, adding "We are responding with dialogue and a shield, and with the shield we are saying we will further improve our deterrent."[56] Kim also said that if nuclear talks were to resume in earnest, North Korea would insist that the nuclear

---

53 "North Korea demands U.S. lift financial restrictions before it will dismantle nuclear program," *Associated Press,* December 20, 2006.

54 Edward Cody, "Nuclear Talks With N. Korea End in Failure, Six-Party Process Thrown Into Doubt," *Washington Post*, December 23, 2006, p. A12.

55 Mitchell Landsberg, "N. Korea nuclear talks end with no resolution," *Los Angeles Times*, December 23, 2006.

56 Ibid.

weapons issue be put aside until others are dealt with first. And North Korea's stand may signal the increased difficulty in persuading Pyongyang to give up its nuclear programs, now that it tested a nuclear device and declared itself a nuclear weapons state.

U.S.-China Cooperation: Key to the Nuclear Issue

The North Korean nuclear test, which China publicly warned against, became a momentum for Chinese policy-makers to put more pressure on Pyongyang. China's policy toward the DPRK has changed, thus pushing Pyongyang harder than before to keep its commitments in the 9.19 joint agreement. The second session of the fifth round of the six-party talks was spent in one-on-one discussions, particularly between the U.S. and North Korea, while others involved either country or China. Gradual changes in Washington's policy towards Pyongyang and North Korea's nuclear test had contributed to a shift in relations between the U.S. and China. Washington and Beijing have moved closer than before. U.S.-China cooperation is essential to a peaceful resolution of North Korea's nuclear issue.

As discussed above, three significant characteristics of this round of the six-party talks should be pointed out. First, the DPRK still holds a key to a resolution of its second nuclear crisis. Second, the U.S. has changed its hard-line policy toward North Korea since the November 2006 elections. Third, U.S.-China strategic cooperation, which is essential to the eventual dismantlement of North Korea's nuclear weapons programs, became manifest.

Hill-Kim Meetings in Berlin

Assistant Secretary Christopher Hill and Vice Minister Kim Kye-gwan held three-day talks in Berlin on January 16-18, 2007 to discuss the nuclear and the BDA issues. The talks made mutually satisfying progress for the next round of the six-party talks. Hill said, "It was a substantive discussion," and "The proof of the pudding will be when we all sit down together in the six-party negotiations."[57] He said on January 20 in Tokyo that the U.S. and the DPRK agreed to hold the next round of the six-party talks and also agreed to the U.S.-DPRK bilateral working-level talks to discuss the financial issue.[58] The agreement would mark a shift in North Korea's long-held position. Previously, Pyongyang said it would not discuss nuclear disarmament unless the U.S. first lifted financial restrictions on North Korea. The U.S. was reportedly considering the release of some legitimate funds from the $24 million frozen in North Korean accounts. The Macau government has control over the North Korean

57 Mark Landler and Thom Shanker, "North Korea and U.S. Envoys Meet in Berlin, *New York Times,* January 18, 2007.

58 Hans Greimel, "U.S., N. Korea Agree to Hold Nuke Talks," *Associate Press*, January 20, 2007.

accounts. U.S. officials indicated that if the Treasury Department closed its BDA investigation, some of the $24 million could be released by Macau.[59]

The U.S.-DPRK bilateral talks in Berlin were significant, long overdue, and the first outside the framework of six-party nuclear talks in Beijing. Hill made an important comment: If the DPRK abandons its nuclear programs, the United States is willing to engage in a bilateral process to establish normal relations. Hill said, "We are prepared to go on that road and to really offer North Korea a hand as it moves along the road."[60] Hill's comments signaled, for the first time, that the U.S. would engage in direct bilateral talks with the DPRK, as Pyongyang has long demanded.

A spokesman for the DPRK Foreign Ministry also said that the bilateral talks in Berlin were held "in a positive and sincere atmosphere and a certain agreement was reached there." He continued, "We paid attention to the direct dialogue held by the DPRK and the U.S. in a bid to settle knotty problems in resolving the nuclear issue."[61]

Deal or No Deal?

Hill suggested those roadblocks to the six-party talks would be cleared after his consultations in Berlin with Kim. Kim said in Beijing on January 23, 2007 that he was "satisfied" with the Berlin talks. He also said he had gotten "positive impressions" from Hill that Washington could change its stance toward the North.[62] What did the U.S. and the DPRK agree on? Based on reports and diplomatic sources, some of their agreements appeared as follows:

First, Hill and Kim agreed in principle to hold "parallel talks" on the next round of the six-party talks and the second round of U.S.-DPRK financial working group talks.[63] The DPRK signaled a change in its stance on the next round of six-party talks. The DPRK previously maintained that it would not discuss the nuclear issue unless the U.S. first lifted its financial restrictions on North Korean accounts at BDA.

Second, Hill and Kim appeared to make a deal in which the DPRK would freeze operations at a nuclear reactor at Yongbyon and allow on-site monitoring by the International Atomic Energy Agency as the first steps to abandoning its nuclear program in exchange for energy aid, releasing legitimate funds of the frozen $24 million in North Korean accounts, and economic aid.[64] It was reported that the U.S. gave a positive response to North Korea's demand that Washington consider

---

59 Carol Giacomo, "U.S.said considering release of some N.Korea funds, *Reuters*, January 16, 2007; Louis Charbonneau, "U.S. and N.Korea pursue nuclear talks," *Reuters*, January 17, 2007.

60 Glenn Kessler, "U.S. Open to Bilateral Talks on Ties With N. Korea," *Washington Post*, January 18, 2007, p. A18.

61 "Spokesman for DPRK Foreign Ministry on Results of DPRK-U.S.Talks," KCNA, January 19, 2007. online at <http://www.kcna.co.jp/index-e.htm> (January 20, 2007).

62 "N. Korea Envoy Notes Movement in U.S. Stance on Nuclear Weapons,"*Associated Press,* January 23, 2007.

63 Kyodo News, January 22, 2007.

64 *Chosun Ilbo*, Janaury 22, 2007.

transforming the Korean armistice into a peace treaty as soon as it starts implementing the initial measures.[65]

To what extent did the U.S. and North Korea agree on the BDA issue? Vice Minister Kim expressed satisfaction with the results of Hill-Kim meetings in Berlin, where they satisfactorily found a formula for resolving the BDA issue. South Korea asked the U.S. to consider unfreezing at least five of North Korea's 50 accounts with the BDA. The U.S. official said Washington agreed that the five accounts were not evidently related to the drug trade, money laundering, U.S. dollar counterfeiting, or other illicit activities. Since they were believed to have been raised through normal trade deals, the U.S. considered unfreezing them to provide North Korea with a chance to start dismantling its nuclear program.[66]

The U.S. reportedly would unfreeze legal accounts to resolve the BDA issue if the DPRK promises to punish those responsible for counterfeiting U.S. dollar bills and money laundering and stop illicit activities in the future. Washington could distinguish between legal and illegal accounts to resolve the BDA issue. At first, Treasury Department officials ruled out making the distinction as the U.S. said that the illegal accounts were frozen as part of law enforcement. However, the Treasury reportedly eased its position because the Bush administration wanted to make progress at the next round of six-party talks. ROK Foreign Minister Song Min Soon said that North Korea showed a willingness to take steps toward dismantling its nuclear program after the United States offered "concessions" during the Hill-Kim bilateral talks.[67] Since North Korea and the U.S. showed their willingness to agree on a road map to implement the 9.19 agreement, as of writing of this chapter, substantial progress could be made at the next round of the six-party talks to be held on 8 February 2007.

The next round of the six-party talks will make a significant progress if North Korea accepts U.S. proposals. In short, the U.S. and the DPRK agreed to hold parallel talks on the nuclear issue and U.S. financial restrictions issue. The second U.S.-DPRK financial working-level talks on the BDA issue were held in Beijing on 30-31 January 2007. Though the talks ended without any breakthroughs, the BDA issue is expected to be resolved soon.

What if the BDA issue is not resolved soon? What will and can Pyongyang do? The six-party talks went into recess without setting a date for the next round. But the six participants pledged to "reconvene at the earliest opportunity." North Korea made it clear that it would stay away from future six-party talks unless the U.S. lifted its financial restrictions on North Korea. If the U.S. is unwilling to resolve the BDA issue, as in the past, Pyongyang will raise the stakes and do everything from provocations to more missile and nuclear tests in order to get what it wants. The DPRK will increase tension once again, probably by testing the second nuclear test.

65 Ibid.

66 "Seoul 'Asked U.S. to Unfreeze N.Korean Accounts,'" *Chosun Ilbo*, January 23, 2007.

67 "In North Korea talks, U.S. made 'concessions,'" Associated Press, January 24, 2007.

North Korea said that the future of the six-party talks will depend on whether the U.S. is willing to change its hostile policy towards Pyongyang.

Now is the time for the U.S. and the DPRK to build up mutual trust by making concessions about the BDA issue. Since North Korea regards the lifting of the financial sanctions against North Korea as the key to a resolution of the North Korea's nuclear issue, the U.S. should seriously consider lifting the financial restrictions and unfreezing some of the frozen $24 million to save North Korean face.

The Author's Three-phase Roadmap for Denuclearizing Korean Peninsula

The DPRK and other five participants at the next round of the six-party talks should agree on detailed plans for implementing the September 19 joint agreement. Thus, the author would like to propose a three-phase road map for denuclearizing the Korean peninsula.[68]

Phase 1. Preparation for dismantlement of North Korea's nuclear programs

At the initial stage, the DPRK should make a complete declaration of all its nuclear programs and freeze all nuclear activities in order to make preparations for shutting down, disabling, and eventually dismantling all nuclear weapons and existing nuclear programs. The DPRK must again join the NPT and allow the IAEA's special inspections of its declared nuclear facilities.

The HEU program will be a potentially very hot issue. There are allegations that the DPRK has its own HEU program, but the U.S. has not provided hard evidence supporting the claims. Therefore, the DPRK first must give an account for what has happened to some twenty centrifuges that were reportedly provided by Dr. A.Q. Khan, and then renounce its HEU program since it has denied the existence of the HEU program. Thus, North Korea must declare that it will not have it in the future. If the DPRK renounces its enriched uranium in a verifiable manner, the five other nations should discuss their economic assistance to the DPRK and the LWR issue with North Korea. The ROK should also discuss its promised supply of electricity to the DPRK. The five nations should provide written security guarantees to the DPRK. All six nations should also discuss and agree to verification procedures. These measures should be simultaneously taken without preconditions.

Phase 2. Actual dismantlement of nuclear weapons and facilities

The DPRK should implement all agreements on eventual dismantlement of nuclear weapons and nuclear facilities in a phased manner. The IAEA should inspect and verify dismantled nuclear programs and North Korea's past nuclear activities. The five other nations and the DPRK should agree on a new construction of LWR or

68 For an earlier version of the roadmap, see Tae-Hwan Kwak, "The Six-Party Nuclear Talks and the Korean Peninsula Peace Regime Initiative: A Framework for Implementation," in Tae-Hwan Kwak and Seung-Ho Joo, eds., *The United States and the Korean Peninsula in the 21st Century* (Aldershot, England: Ashgate, 2006), pp.17-19; pp. 26-28.

resumption of the suspended KEDO project at Sinpo under new arrangements. At the same time, U.S.-DPRK normalization and DPRK-Japan normalization talks should proceed. It is desirable that a six-party summit meeting be held at this phase in order to build up mutual confidence among the six-nations. The four-party (the U.S., China and two Koreas) should be held to discuss a peace treaty ending the Korean war.

Phase 3. Conclusion of the "Six-Party's Korean Peninsula Denuclearization Guarantee Agreement"

The end of nuclear dismantlement will be at the third phase, during which the six parties should conclude a denuclearization guarantee agreement in which, institutional and legal arrangements for enforcement measures of the denuclearization on the Korean peninsula should be contained. The six participants must draw up a multilateral nuclear agreement in which North Korea would completely dismantle its nuclear programs in exchange for multilateral security guarantees and economic cooperation. This agreement should be registered with the United Nations Secretariat. The construction of LWR should be under way, and massive economic assistance to North Korea will be given, and U.S.-DPRK and Japan-DPRK normalization agreements will be signed at this phase (see Table 2-1).

In the final analysis, there will be a long and rough road ahead to a peaceful resolution of North Korea's nuclear issue. The U.S. and the DPRK should continue to cooperate through compromises and concessions to achieve the verifiable denuclearization of the Korean peninsula.

Concluding Remarks

The U.S. and the DPRK will make all efforts to compromise their conflicting positions on the BDA and nuclear issues. The BDA issue was a key obstacle to the second session of the fifth round of six-party talks in December 2006. It is desirable that President Bush and Chairman Kim be flexible about the BDA issue. The DPRK reportedly demanded that the U.S. unfreeze some of the frozen $24 million North Korean accounts at BDA. Chairman Kim has shown his flexibility on the BDA issue.

The U.S. and the DPRK need to make a "political" compromise to save their faces. Without making mutual concessions, the BDA issue would not be resolved. Thus, the DPRK will have to take the initiative in resolving the issue just as it admitted that it had previously abducted Japanese citizens. The DPRK should simply admit to engaging in money laundering and it would have to stop engaging in the illicit activities in return for Washington's lifting of the financial restrictions on some of the North Korean accounts at BDA. Should the DPRK decide to atone for its past wrongdoing and take remedial measures, Washington should reciprocate it. These actions need to be simultaneously taken by both sides. With the resolution of the BDA issue, the six-party talks will be activated to take actions for implementing the denuclearization of the Korean Peninsula. Chairman Kim must make a bold decision to commit again to dismantlement of all nuclear weapons programs.

The DPRK and other five nations need to sincerely take the initial steps for building mutual confidence between the U.S. and the DPRK, that will be a firm foundation of peace and stability on the Korean peninsula and in Northeast Asia.

Table 2.1 Three-phase road map for denuclearization of Korean peninsula

Objectives and Goals of Different Stages \ Concerned Parties		North Korea (NK)	International Community (5 Parties +)
Three-phase roadmap for verifiable denuclearization of Korean peninsula	1st Phase: Preparation phase (Disclosure of all nuclear programs, including HEU)	• Shutdown of all nuclear facilities •Abandoning of all nuclear weapons/existing nuclear programs • Return to NPT/ IAEA inspections • Renunciation of HEU • Verification procedures agreement • Inter-Korean Summit meeting	• Fuel oil supply to NK (5 parties) discussed • Economic aid to NK discussed •Verification procedures agreement • Written security guarantees by 5 Parties • South Korea's supply of electricity to NK discussed •LWR provision discussed
	2nd Phase: Dismantlement of nuclear weapons and facilities	• Dismantlement began • IAEA inspections of nuclear facilities • NK's agreement on IAEA inspections on its past nuclear activities •Six-nation summit meeting/ Four party forum	•Resumption of suspended LWR or new construction of LWR •NK-US, NK-Japan normalization talks began •Six-nation summit meeting/ Four party forum

	3rd Phase: Conclusion of Six- Party's Guarantee on Korean peninsula denuclearization	•End of nuclear dismantlement • LWR nuclear reactor construction began •Joint Korean peninsula denuclearization agreement registered with UN Secretariat	• LWR nuclear reactor construction began • Promise of grand economic assistance program to NK • U.S.-NK, Japan-NK normalization agreement signed •Joint Korean peninsula denuclearization agreement registered with UN Secretariat

Appendix 2.1

Joint Statement of the Fourth Round of the Six-Party Talk

Following is a text of the joint statement at the conclusion of the fourth round of Six-Party Talks, as released in Beijing on September 19, 2005 by the Ministry of Foreign Affairs of the People's Republic of China.

Joint Statement of the Fourth Round of Six-Party Talks Beijing 19 September 2005

The Fourth Round of the Six-Party Talks was held in Beijing, China among the People's Republic of China, the Democratic People's Republic of Korea, Japan, the Republic of Korea, the Russian Federation, and the United States of America from July 26th to August 7th, and from September 13th to 19th, 2005.

Mr. Wu Dawei, Vice Minister of Foreign Affairs of the PRC, Mr. Kim Gye Gwan, Vice Minister of Foreign Affairs of the DPRK; Mr. Kenichiro Sasae, Director-General for Asian and Oceanian Affairs, Ministry of Foreign Affairs of Japan; Mr. Song Min-soon, Deputy Minister of Foreign Affairs and Trade of the ROK; Mr. Alexandr Alekseyev, Deputy Minister of Foreign Affairs of the Russian Federation; and Mr. Christopher Hill, Assistant Secretary of State for East Asian and Pacific Affairs of the United States attended the talks as heads of their respective delegations.

Vice Foreign Minister Wu Dawei chaired the talks.

For the cause of peace and stability on the Korean Peninsula and in Northeast Asia at large, the Six Parties held, in the spirit of mutual respect and equality, serious and practical talks concerning the denuclearization of the Korean Peninsula on the basis of the common understanding of the previous three rounds of talks, and agreed, in this context, to the following:

1. The Six Parties unanimously reaffirmed that the goal of the Six-Party Talks is the verifiable denuclearization of the Korean Peninsula in a peaceful manner.

The DPRK committed to abandoning all nuclear weapons and existing nuclear programs and returning, at an early date, to the Treaty on the Non-Proliferation of Nuclear Weapons and to IAEA safeguards.

The United States affirmed that it has no nuclear weapons on the Korean Peninsula and has no intention to attack or invade the DPRK with nuclear or conventional weapons.

The ROK reaffirmed its commitment not to receive or deploy nuclear weapons in accordance with the 1992 Joint Declaration of the Denuclearization of the Korean Peninsula, while affirming that there exist no nuclear weapons within its territory.

The 1992 Joint Declaration of the Denuclearization of the Korean Peninsula should be observed and implemented.

The DPRK stated that it has the right to peaceful uses of nuclear energy. The other parties expressed their respect and agreed to discuss, at an appropriate time, the subject of the provision of light water reactor to the DPRK.

2. The Six Parties undertook, in their relations, to abide by the purposes and principles of the Charter of the United Nations and recognized norms of international relations.

The DPRK and the United States undertook to respect each other's sovereignty, exist peacefully together, and take steps to normalize their relations subject to their respective bilateral policies.

The DPRK and Japan undertook to take steps to normalize their relations in accordance with the Pyongyang Declaration, on the basis of the settlement of unfortunate past and the outstanding issues of concern.

3. The Six Parties undertook to promote economic cooperation in the fields of energy, trade and investment, bilaterally and/or multilaterally.

China, Japan, ROK, Russia and the US stated their willingness to provide energy assistance to the DPRK.

The ROK reaffirmed its proposal of July 12th 2005 concerning the provision of 2 million kilowatts of electric power to the DPRK.

4. The Six Parties committed to joint efforts for lasting peace and stability in Northeast Asia.

The directly related parties will negotiate a permanent peace regime on the Korean Peninsula at an appropriate separate forum.

The Six Parties agreed to explore ways and means for promoting security cooperation in Northeast Asia.

5. The Six Parties agreed to take coordinated steps to implement the aforementioned consensus in a phased manner in line with the principle of "commitment for commitment, action for action".
6. The Six Parties agreed to hold the Fifth Round of the Six-Party Talks in Beijing in early November 2005 at a date to be determined through consultations.

Released on September 19, 2005

Source: U.S. Department of State online at <http://www.state.gov/r/pa/prs/ps/2005/53490.htm>

Chapter 3

The Bush Administration and North Korea's Nuclear Policy

Edward A. Olsen

Ever since the creation of the Democratic People's Republic of Korea (DPRK) in the northern half of the Korean peninsula in 1948, U.S. relations with North Korea have ranged from poor to hostile. Prior to the DPRK's formation, during the very early phases of the 1945-48 time frame, there was at least a hypothetical possibility that the United States could have created positive ties with the northern Koreans before the Soviet Union shaped what became "North Korea." After all, the United States did not enter the divided Korean nation after its liberation from Imperial Japan with a clear vision of what Korea's future should encompass. Nor did Americans harbor any particular animosity or friendliness toward either half of the divided nation.[1] Had the Cold War not spread from Europe to Asia and created U.S.-USSR tensions concerning each's ambitions in and surrounding the Korean peninsula, a plausible case can be made that the United States had reason to be as interested in—if not more interested in—northern Korea compared to southern Korea because of the northern region's industrial base resulting from Japanese colonial policies. Such possibilities were eliminated by the Cold War's enveloping Korea and magnified by the United States' strategic commitments in occupied Japan and the Soviet Union's inability to play a serious role in postwar Japan due to U.S. policies. In short, a dynamic rapidly evolved around Korea that made the peninsula what Gregory Henderson insightfully characterized as the "vortex" of Asian geopolitics.[2] That dynamic escalated as a result of overconfident North Korean misperceptions of American ambivalent commitments to South Korea shaped by post-World War II American popular desires to bring U.S. forces home. This led to periodic inter-Korean cross-border tensions and then large scale North Korean aggression that yielded the Korea War which, despite its name, rapidly became an internationalized war and a major turning point for U.S. policy toward both the Korean peninsula and the larger Cold War.[3]

---

1 The author explored that situation in greater detail in Chapter 4, "Liberation and Division," of *Korea, The Divided Nation* (Westport: Praeger Security International, 2005).

2 Gregory Henderson, *Korea: The Politics of the Vortex* (Cambridge: Harvard University Press, 1968).

3 For a spectrum of analyses of the Korean War and its Cold War context, see David Rees, *Korea: The Limited War* (New York: St. Martin's, 1964); J. Lawton Collins, *War in Peacetime: The History and Lessons of Korea* (Boston: Houghton Mifflin Co., 1969); Joseph

Adversarial Relations

Because of the legacy of the Korean War the United States' adversarial relations with North Korea became a core element of overall U.S. foreign and defense policy regionally and globally. A generation of Americans and North Koreans has grown accustomed to this factor in each country's approach to the other. For better or worse, and usually for the worse, this factor looms much larger in North Korean consciousness than it does among Americans because the United States for more than half a century has posed a range of major threats to the DPRK either directly or by backing the Republic of Korea (ROK) strategically—including a major long-term deployment of U.S. armed forces in South Korea. Conversely, despite American perceptions of North Korea as posing a threat to U.S. national interests, that threat is only one of many worldwide threats perceived by Americans throughout the Cold War and post-Cold War eras. More important, North Korea's ability to threaten the United States' territorial homeland has routinely been overshadowed by other players in the international system. That situation began to change for the worse when North Korea adjusted to post-Cold War circumstances by exploring its nuclear weapons option as a way to reinforce the DPRK's ability to adapt to the elimination of the Soviet Union as an ally and the geopolitical uncertainties inherent in the PRC's integration within the global economic market place.

U.S. policy toward the DPRK's nuclear agenda has been the focus of numerous articles and books.[4] During the formative stage of North Korea's nuclear weaponry ambitions in the latter years of the Cold War the United States clearly was attentive

---

C. Goulden, *Korea: The Untold Story of the War* (New York: McGraw-Hill, 1982); Jon Halliday and Bruce Cumings, *Korea: The Unknown War* (New York: Pantheon, 1988); James Cotton and Ian Neary, eds., *The Korean War in History* (Atlantic Highlands, NJ: Humanities Press International, 1989); and William Stueck, *The Korea War: An International History* (Princeton: Princeton University Press, 1995). In addition, for insights into U.S. policies toward the Korean War, see: Rosemary A. Foot, *The Wrong War: American Policy and the Dimensions of the Korean Conflict* (Ithaca: Cornell University Press, 1985); and James I. Matray, *The Reluctant Crusade: American Foreign Policy in Korea, 1941-1950* (New York: St. Martin's Press, 1997).

4 Prominent examples of such books, include: Michael J. Mazaar, *North Korea and the Bomb: A Case Study in Nonproliferation* (New York: St. Martin's Press, 1995); Peter Hayes and Young Whan Kihl, eds., *Peace and Security in Northeast Asia: The Nuclear Issue and the Korean Peninsula* (Armonk: M.E. Sharpe, 1996); Leon V. Sigal, *Disarming Strangers: Nuclear Diplomacy with North Korea* (Princeton: Princeton University Press, 1999); Victor D. Cha and David C. Kang, *Nuclear North Korea, A Debate on Engagement Strategies* (New York: Columbia University Press, 2003); Joel Wit, Robert L. Gallucci, and Daniel B. Poneman, *Going Critical: The First North Korean Nuclear Crisis* (Washington, DC: Brookings Institution Press, 2003); Michael O'Hanlon and Mike Mochizuki, *Crisis on the Korean Peninsula: How to Deal With a Nuclear North Korea* (New York: McGraw-Hill, 2003); Gavan McCormack, *Target North Korea: Pushing North Korea to the Brink of Nuclear Catastrophe* (New York: Nation Books, 2004); William C. Triplett III, *Rogue State: How a Nuclear North Korea Threatens America* (Washington, DC: Regnery Publishing, 2004); and Gordon G. Chang, *Nuclear Showdown: North Korea Takes on the World* (New York: Random House, 2006).

to what was emerging in the DPRK, but that facet of U.S. overall foreign and defense policy did not have a particularly high profile. In part that was because the credibility of North Korea's nuclear ambitions, despite coming close to war in the mid-1990s, did not always seem very serious and the blend of facts and rumors the United States was learning about with regard to what was transpiring in North Korea tended to justify such judgments. More importantly the United States was preoccupied by how the Cold War to post-Cold War transition was evolving globally and assessing its impact on other potentially rising threats. Much of that was altered profoundly a decade into the post-Cold War period by the geopolitical trauma experienced by the American public and U.S. leaders as a result of the 9/11 radical Islamic terrorist attacks on the World Trade Center and the Pentagon. The sense of vulnerability in the United States escalated tremendously as did the patriotic fervor surrounding the "Global War on Terrorism" (GWOT). Rightly or wrongly, North Korea found itself caught up in American strategic anxieties as a result of President George W. Bush's inclusion of the DPRK as part of the infamous "Axis of Evil" in Bush's 2002 State of the Union speech.[5] In part because the axis of evil phrase had roots in a neo-conservative writer, David Frum, formerly with the hawkish neo-con magazine, *The Weekly Standard*,[6] North Korea predictably reacted very critically.[7] In this context North Korea's long-standing fears that the United States was committed to toppling the DPRK grew as the United States increasingly saw North Korea as an anachronistically rogue remnant of the former Cold War. The expansion of the "GWOT" in the form of a corollary conflict in Iraq—one of the infamous "Axis"—further alarmed North Korean leaders because of perceptions that U.S. advocates of DPRK "regime change" had growing influence in Washington.[8]

These growing tensions between the United States and North Korea centered on the DPRK nuclear weapons agenda that was made more ominous to post-9/11 Americans by the possibility that the Kim Jong-Il government might try to take advantage of the United States' preoccupation with Islamic terrorists and the ways in which conventional U.S. armed forces might be stretched too thin to be readily used in Korea. Conversely such U.S. concerns and restraints made the U.S. threat to the DPRK more ominous to North Koreans because Americans might be tempted to apply the Bush Doctrine's preemptive paradigm to the Korean situation. Clearly North Korea feared the United States might apply the Iraq approach to the DPRK, might

5 David E. Sanger, "In Speech, Bush Calls Iraq, Iran, and North Korea 'an Axis of Evil,'" *New York Times*, January 30, 2002.

6 For coverage of its roots, see the Lexington column, "Paul Wolfowitz, Velociraptor," *The Economist*, February 9, 2002. For an assessment of The Weekly Standard's role within American conservatism, see George Will's characterization of it as "spectacularly misnamed radicalism" in his column, "Transformation's Toll," *Washington Post*, July 18, 2006.

7 John Ward Anderson, "Angry Denials from the 'Axis,'" *Washington Post*, January 31, 2002; and David R. Sands, "North Korea Assails 'Axis' Label," *Washington Times*, February 1, 2002.

8 For pointed examples, see: John McCain, "Rogue State Rollback," *Weekly Standard*, January 20, 2003; Henry S. Rowen, "Kim Jong-Il Must Go," *Policy Review* (Hoover Institution), October/November 2003; and Nicholas Eberstadt, "Tear Down This Tyranny," *Weekly Standard*, November 29, 2004.

push for internal regime change, or might go far beyond either of these alternatives by eliminating the DPRK nuclear option through elimination of the DPRK via an obliterating military attack. As such issues loomed on the horizon, it became more obvious to all observers—and certainly to officials in North Korea—that the United States was increasingly pursuing policies that differed from its South Korean ally. Solid examples of such U.S. policies were on the human rights and economic fronts. Congressional passage of the 2004 "North Korean Human Rights Act"[9] and the ways that legislation stirred up the debate over an appropriate U.S. policy toward North Korea[10] compounded the United States' problems on the nuclear weapons issue. The 2005 legal charges of North Korean counterfeiting of U.S. currency[11] which North Korea controversially resisted in a manner that some critics of U.S. policy accepted,[12] added another layer of complications that detracted from progress in the six-party talks tasked with engaging North Korea in a productive dialogue.[13] While the human rights and counterfeit issues had factual merit, neither helped U.S.-ROK cooperation on policies toward the DPRK and inter-Korean relations. Neither U.S. goals on its North Korean human rights agenda nor U.S. objectives in controlling illicit financial activities that can prop up a weakened North Korean regime should be permitted to get in the way of more constructive U.S. policies designed to resolve the pressing nuclear issues.

Those problems have been compounded by two other contextual developments. One clearly is the responsibility of North Korea—namely the DPRK's attempt to add to its deterrence capabilities by carrying out missile launch tests on the fourth

---

9 See H.R. 4011 – North Korean Human Rights Act of 2004.

10 For coverage of the Act's impact on U.S. policy, see Karin J. Lee, "The North Korean Human Rights Act and other Congressional Agendas," *Nautilus Institute Policy Forum Online*, October 7, 2004, available online at <www.nautilus.org> and Jacques G. Boetcher, "The North Korean Human Rights Act of 2004: More Rhetoric or a New U.S. Policy Statement," *North Korean Review*, Spring 2006.

11 See Glenn R. Simpson, Gordon Fairclough, and Jay Solomon, "U.S. Probes Banks," *Wall Street Journal*, September 9, 2005; Jay Solomon, "U.S. Files Charges In North Korean Counterfeit Probe," *Wall Street Journal*, October 12, 2005; Bill Gertz, "U.S. Accuses North Korea of $100 Bill Counterfeiting," *Washington Times*, October 12, 2005; and Tim Johnson (Knight Ridder), "U.S. Accuses N. Korea of Money Scam," *Monterey Herald*, October 29, 2005.

12 For coverage of this issue's evolution, see Ryu Jin, "NK Sending Positive Signals To Outside World, Pyongyang Likely to Come Out to Address Financial Troubles With US," *Korea Times*, February 13, 2006; Annie I. Bang, "N. Korea Seen Willing to Resolve Counterfeiting Issue: Lawmakers," *Korea Herald*, February 13, 2006; and Josh Meyers, "Squeeze On North Korea's Money Supply Yields Results," *Los Angeles Times*, November 2, 2006.

13 For evaluations of that impact, see Martin Fackler, "North Korea Counterfeiting Complicates Nuclear Crisis," *New York Times*, January 29, 2006; Lee Kwang-ho, "North Korea-U.S. Standoff Over Financial Sanctions," *Vantage Point*, March 2006; Kwak Tae-hwan, "The U.S. Financial Sanctions and Six-Party Talks," *Vantage Point*, April 2006; Bill Gertz, U.S. Sanctions Cost North Korea Millions," *Washington Times*, June 16, 2006; and Larry Niksch, "North Korea's Nuclear Weapons Development and Diplomacy," CRS Report for Congress, October 25, 2006.

of July, 2006. It is obvious that North Korea's selection of that date was designed to send a message to Americans on that U.S. holiday about the independence of Korea. Although the missile tests were not as technologically successful as Pyongyang no doubt hoped they would be, they nonetheless were geopolitically successful in terms of conveying North Korea's commitment to be assertive.[14] That form of success was underscored by North Korea's willingness to carry out the tests despite preemptive warnings from the United States designed to get the DPRK to halt the missile tests.[15] How much impact those tests, and potential additional tests, will have on U.S. policy toward North Korea over the long run remains to be seen. Based on both sides' policies toward each other, it is safe to assume that the United States and North Korea probably will approach this issue from a brinkmanship perspective, pushing each other's envelope.

The second contextual development differs greatly in that North Korea bears no responsibility for it. This contextual shift derives from the United States' odd mixture of a hard line approach to Iran's nuclear weapons agenda[16] and flexibly accepting approach toward India becoming a member of the so-called global nuclear club.[17] The contrast between the United States' policies toward these two regional powers with rising international aspirations is striking and subject to considerable debate about their soundness. More important for present purposes is the glaring disparity between the U.S.-India nuclear dynamic and the United States' opposition to today's North Korean nuclear ambitions and the United States' past foreclosure of a nuclear option for South Korea.[18] In addition to adding to the alienation evident in North Korea's posture vis-à-vis the United States, it also suggests some hypocrisy

---

14 For assessments of the missile test's impact, see Robert Marquand, "Korea's Missile Salvo To World," *Christian Science Monitor*, July 6, 2006; Lauren Etler, "North Korea's July 4th Display: How Big A Threat?" *Wall Street Journal*, July 8-9, 2006; "Kim Jong Il Goes Ballistic," *The Economist*, July 8, 2006; and Lee Kwang-ho, "Will North Korea's Missile Gamble Succeed?" *Vantage Point*, July 2006.

15 For prominent examples, see Demetri Sevastopulo and Anne Fifield, "U.S. Warns On Readiness of North Korean Missiles," *London Financial Times*, June 16, 2006; and Ashton B. Carter and William J. Perry, "If Necessary, Strike and Destroy; North Korea Cannot Be Allowed to Test This Missile," *Washington Post*, June 22, 2006.

16 For useful coverage of that agenda and the United States' responses, see "Iran's Nuclear Secrets; Coaxing Iran To Come Clean," *The Economist*, November 29, 2003; Thomas Omestad, "Iran and The Bomb," *U.S. News & World Report*, November 22, 2004; Scott Peterson, "Iran Bids to Redefine Nuclear Limits," *Christian Science Monitor*, September 19, 2005; and "Special Report: Iran's Nuclear Programme; When The Soft Talk Has To Stop," *The Economist*, January 14, 2006.

17 For coverage of the U.S.-India nuclear deal and the controversy it aroused, see "India and America; Joining the Nuclear Family," *The Economist*, March 4, 2006; Dana R. Dillon and Baker Spring, "Nuclear India and The Non-Proliferation Treaty," The Heritage Foundation *Backgrounder*, No. 1935, May 18, 2006; Paul Richter, "Bush's India Plan At Risk," *Los Angeles Times*, June 3, 2006; "America's Nuclear Deal With India; From Bad To Worse," *The Economist*, July 22, 2006; and Peter Baker, "Bush Signs India Nuclear Law," *Washington Post*, December 19, 2006.

18 For background on what the United States did during the Park Chung-hee administration, see Chapter 20, "South Korea and Nuclear Weapons," in Selig S. Harrison,

regarding the criteria used by U.S. policy makers. Assuming the cooperative U.S.-India nuclear paradigm persists, one can readily imagine a future united Korea—which will inherit the nuclear know how of today's North Korea—making a plausible case that today's Bush doctrine regarding a nuclear-armed India with its heritage of Cold War non-alignment should be applied to a united Korea built on the legacy of over half a century of close U.S.-ROK alliance relations. It is easy to imagine Koreans living in and leading that united Korea wanting to be treated like India and encouraged to have nuclear weapons.

Contextual Legacies

Consequently, against this background and in order to set the stage for an assessment of how U.S. policy in the Bush administration toward the North Korean nuclear situation can be improved in the remainder of its second term in ways that will improve its legacy for successor administrations, it is necessary to succinctly examine why and how U.S. policy toward North Korea within the context of U.S. policy toward the entire Korean peninsula and its regional setting evolved. United States policy toward the divided Korean nation has been convoluted from the start of Korea's division in the early days of the Cold War. American policy makers in that formative period were ill prepared for dealing with Korea, naively accepted Korea's Japanese absorption from 1910 to 1945 into what became an enemy nation in World War II, knew little about its lengthy history or national aspirations after being liberated from Japanese Imperialism, and pursued profoundly ambiguous policies. Those policies could not articulate what U.S. national interests vis-à-vis the entire Korean peninsula were and paid virtually no attention to the spectrum of Korean views about Korean national interests. In retrospect the poor quality of post-World War II U.S. policy toward Korea was not surprising, given the United States' history of flawed decisions toward Korea in the nineteenth and early twentieth centuries. In contrast, the Soviet Union's policies toward the Korean nation in the formative phase of the Cold War were better planned in terms of Soviet Marxist interests, drew on long-standing Russian ambitions regarding Korea, and were pursued in ways that were very attentive to Moscow's perception of Korean sensitivity about Korean national interests.[19]

A major consequence of those circumstances was the close ties between the USSR and its North Korean protégé which was largely an offshoot despite some dissenting factions that shaped both the origins of the Korean War and its legacy for the Asian theater of the Cold War versus the more ambivalent U.S.-South Korean relationships prior to the Korean War that were solidified the hard way during that conflict in a manner that also shaped the evolution of the Cold War in Asia. In short,

---

Korean Endgame: A Strategy for Reunification and U.S. Disengagement (Princeton: Princeton University Press, 2002).

19 For in-depth coverage of that history and the United States' role in it, see Bruce Cumings, *Korea's Place in the Sun: A Modern History* (New York: W.W. Norton, 1997); and Don Oberdorfer, *The Two Koreas: A Contemporary History* (Reading, MA: Addison-Wesley, 1997).

U.S. policy toward the Korean nation divided into two rival states became far firmer but still very convoluted. The U.S.-ROK alliance relationship after the Korean War truce settled into a pattern of close mentor-client state bonds while the U.S.-DPRK adversarial relationship became entrenched, bitter, and mutually ill-informed. Even if one accepts the perceptions of critics of the U.S.-ROK relationship who saw it as too close in terms of the United States being bossy and South Korea being servile, the U.S.-ROK relationship was far better than the U.S.-DPRK relationship. In the context of this historical background the United States' policies, past and present, toward the two Korean states expounding each's quest for security, economic prosperity, and political self-determination, as well as the divided Korean nation's aspirations for reunification, have been and are troubled by U.S. problems in dealing with how those issues and aspirations interact with each other. In short, they do not mesh well with the evolution of U.S. national interests toward Korea and its Asian regional setting.

The United States' bilateral relations with each half of Korea have, as stated, had an ambiguous record. U.S.-DPRK relations have, for obvious reasons, been the most tenuous. On the official diplomatic level they do not exist in terms of mutual recognition. Ever since the ROK's formal creation in 1948 the United States has recognized it as Korea's only legitimate government. While the United States clearly acknowledges the DPRK's existence, for decades has dealt with the threats North Korea poses, and throughout this period has had formal military-to-military and—more recently—diplomatic contacts with DPRK representatives, those contacts have at best been artificially formal and at worst were bitterly contentious. In that vein the DPRK's two state leaders, the father and son "dynasty" of Kim Il-Sung and Kim Jong-Il,[20] have received the de facto but not de jure recognition of a succession of U.S. administrations. None of this was helped, in the context of North Korea's nuclear agenda, by bleak American perceptions of North Korea's less than pragmatic approaches to its domestic problems spawned by food shortages,[21] economic malfeasance,[22] and dismal human rights record.[23] Had it not been for the

20 For assessments of its dynastic succession issues, see Tai Sung An, *North Korea in Transition: From Dictatorship to Dynasty* (Westport: Greenwood Press, 1983); and Young Whan Kihl, "Staying Power of the Socialist 'Hermit Kingdom,'" in Young Whan Kihl and Hong Nack Kim, eds., *North Korea, The Politics of Regime Survival* (Armonk: M.E. Sharpe, 2006).

21 For examples of analyses of North Korea's famines that influenced American perceptions, see Andrew Natsios, *The Politics of Famine in North Korea* (Washington, DC: U.S. Institute of Peace, 1999); and his *The Great North Korean Famine; Famine, Politics, and Foreign Policy* (Washington, DC: U.S. Institute of Peace, 2002).

22 For examples of analyses of North Korea's economic policies that influenced American perceptions, see John Pomfret, "Reforms Turn Disastrous for North Koreans," *Washington Post*, January 27, 2003; Marcus Noland, *Korea After Kim Jong-Il* (Washington, DC: Institute for International Economics, January 2004); and Mark E. Manyin, "Foreign Assistance to North Korea," *CRS Report* (Washington, DC: Congressional Research Service, 2005).

23 For a prime example of a book that had major influence on global opinion of North Korea's human rights and socioeconomic situation which helped to shape U.S. perceptions of

mixture of U.S. post-9/11 policies and North Korea's blend of provocative nuclear brinkmanship strategy with a defensive response to the Bush administration's possible pursuit of regime change in Pyongyang, it is unlikely that the DPRK would have occupied such a prominent niche in the United States' worldwide efforts to cope with terrorists and states that either support terrorist causes materially or behave in a manner that earns them the label "rogue" state. Consequently, the issue of how to deal with North Korea, especially on the nuclear front, became a major concern for the G.W. Bush administration and the focus of many analytical studies about what the Bush administration should do.[24]

Overall U.S. relations with South Korea have been far better than with North Korea. As South Korea's democracy evolved in generally positive ways,[25] its economy matured in ways transforming the ROK from a third world state to a dynamic player in global capitalism,[26] and its international standing grew in keeping with these political and economic trends,[27] the U.S.-ROK relationship far overshadowed anything that existed between the United States and North Korea. Despite those many positive aspects of U.S.-ROK relations, especially in contrast to North Korea, as the ROK matured in diverse ways South Korean attitudes toward the United States and American leadership within the international system also evolved in ways that sometimes did not conform to what U.S. leaders thought a South Korean ally should

---

those issues, see Norbert Vollertsen, *Inside North Korea: A Diary of A Mad Place* (New York: Encounter Books, 2003). See also Hazel Smith, *Hungry for Peace; International Security, Humanitarian Assistance, and Social Change in North Korea* (Washington, DC: U.S. Institute of Peace, 2005).

24 For a cross-section of such studies, see Morton I. Abromowitz and James T. Laney, "Meeting the North Korean Challenge" (New York: Council on Foreign Relations / Independent Task Force Report, 2003); James T. Laney and Jason T. Shaplen, "How to Deal with North Korea," *Foreign Affairs* (March/April 2003); Jonathan D. Pollack, "The United States, North Korea, and the End of the Agreed Framework," *Naval War College Review* (Summer 2003); Michael O'Hanlon and Mike Mochizuki (op. cit./2003); Gavan McCormack (op. cit./2004); and Jasper Becker, *Rogue State: The Continuing Threat of North Korea* (New York: Oxford University Press, 2005).

25 For useful surveys of that evolution, see Manwoo Lee, *The Odyssey of Korean Democracy* (New York: Praeger, 1990); John Kie-Chiang Oh, *Korean Politics* (Ithaca: Cornell University Press,, 1999); and Young Whan Kihl, *Transforming Korean Politics: Democracy, Reform, and Culture* (Armonk: M.E. Sharpe, 2005).

26 For useful surveys of that maturation, see Jon Woronoff, *Korea's Economy: Man-Made Miracle* (Seoul: Si-sa-yong-o-sa Publishers, 1983); Alice H. Amsden, *Asia's Next Giant: South Korea and Late Industrialization* (New York: Oxford University Press, 1989); Byung-Nak Song, *The Rise of The Korean Economy* (Oxford; Oxford University Press, 1990); and Cho Soon, *The Dynamics of Korean Economic Development* (Washington, DC: Institute for International Economics, 1994).

27 For useful surveys of South Korea's growing international roles, see Ahn Byung-joon, *South Korea's International Relations: Quest for Security, Prosperity, and Unification*, New York: The Asia Society, 1991; Young Whan Kihl, ed., *Korea and the World; Beyond the Cold War*, Boulder: Westview Press, 1994; and Samuel S. Kim, ed., *Korea's Globalization*, New York: Cambridge University Press, 2000.

be.[28] Those tensions caused many observers in the United States to raise questions about the long-term prospects for the U.S.-ROK alliance and its implications for overall U.S.-Korean relations.[29] As part of that evolving strategic debate about the alliance scholars at the Cato Institute advocated a U.S.-ROK "amicable divorce."[30] Lest that perception be unfairly ascribed only to a libertarian perspective, a prominent mainstream Asia specialist from Vanderbilt University characterized South Korea as a "feckless ally"[31] and a neo-con analyst of Asian affairs from the American Enterprise Institute labeled South Korea a "runaway ally."[32] The United States' approach to the alliance with South Korea—when coupled with broader American strategic priorities in pursuit of greater geopolitical flexibility, burden sharing with affluent allies, and periodic controversies over the behavior of U.S armed forces that raised questions about the legitimacy of evolving status of forces legal agreements—caused many South Koreans to question the nature of the alliance.[33] Such South Korean attitudes were exemplified at a very high level by South Korean Foreign Minister Song Min-soon's characterization of the United States under George W. Bush as "the most warlike country in history."[34]

Faced with such issues within the U.S.-ROK security relationship as the two allies attempted to develop a coherent joint approach toward North Korea, it was increasingly evident that the G.W. Bush and Roh Moo-hyun administrations were not closely coordinating their policies.[35] This has been most obvious on those occasions

28 For insights into such rifts, see Seung-Hwan Kim, "Anti-Americanism in Korea," *Washington Quarterly*, Winter 2002-03; and David I. Steinberg, ed., *Korean Attitudes Toward the United States, Changing Dynamics* (Armonk: M.E. Sharpe, 2005).

29 For example, see Donald W. Boose, Jr., ed., *Recalibrating the U.S.-Republic of Korea Alliance* (Carlisle: U.S. Army War College, 2003); and Norman D. Levin, *Do The Ties Still Bind? The U.S.-ROK Relationship After 9/11* (Santa Monica: RAND, 2004).

30 See Chapter Five, "Time for an Amicable Divorce" in Ted Galen Carpenter and Doug Bandow, *The Korean Conundrum; America's Troubled Relations with North and South Korea* (New York: Palgrave Macmillan, 2004). See also Doug Bandow, "Allies or Ingrates? The U.S. gets the worst of its relationship with South Korea," *The American Conservative*, June 20, 2005.

31 James E. Auer (from Vanderbilt) and Robyn Lim, "Japan: America's New South Korea?" *Current History*, September 2004.

32 Nicholas Eberstadt, "Tear Down This Tyranny," *The Weekly Standard*, November 29, 2004.

33 For coverage of such issues, see Jung Sung-ki, "Korea-U.S. Military Alliance Turns Sour," *Korea Times*, April 12, 2005; Joo Sang-min, "USFK Incidents Prompt Korean Protest," *Korea Herald*, July 18, 2005; Jin Dae-woong, "'Strategic flexibility' New Credo for USFK," *Korea Herald*, January 21, 2006; Ryu Jin, "Korea-U.S. Alliance Cracking," *Korea Times*, July 21, 2006; and "South Korea and America, Awkward Bedfellows," *Economist*, September 9, 2006.

34 Quoted in Robert Marquand, "South Korea's Beef with U.S. Business," *Christian Science Monitor*, December 14, 2006 (with correction on December 19, 2006).

35 For coverage of those frictions, see Howard W. French, "U.S. Approach on N. Korea Strains Alliances in Asia," *New York Times*, February 4, 2003; and Robert Thompson and Richard Lloyd Parry, "'Don't Go Too Far' South Korean Leader Tells Bush," *London Times*, March 5, 2003.

when President Roh publicly urged the United States to moderate its positions on North Korean issues.[36] As a result some pointed U.S. critics of South Korea's policies emerged in the U.S. media sympathetic to such perspectives.[37] Such attitudes also emerged within high level U.S. policy-making circles, causing a prominent American progressive critic of that approach to identify them and label them "An 'Axis of Evil' within the administration."[38] In short, a more hawkish approach toward North Korea and its nuclear agenda became dominant in U.S. policy which its advocates assumed would be productive[39] and would encourage South Korea to adapt its policies toward North Korea to conform with the posture of the ROK's U.S. ally.

Policy Soundness

Against that background, has the United States' policy toward North Korea in recent times been effective? The blunt answer is an ambiguous one: not as much as its advocates think it is. The United States has not devised the means on its own to induce North Korea to truly abandon its nuclear weapons agenda. The results that have been accomplished undoubtedly were motivated by U.S. interests in containing DPRK nuclear ambitions, but the decisive actions were the result of China and South Korea engaging North Korea in innovative negotiations both at the six-party talks and—arguably more important—in other meetings focused on socioeconomic means to facilitate Korean national reconciliation.

The North Korean nuclear agreement brokered by China in September 2005 at the six-party talks in Beijing represented progress for the United States' ambition of blocking the DPRK's nuclear weapons ambitions. As important as the agreement to halt North Korea's nuclear weapons programs, return to the Nuclear Nonproliferation Treaty (NPT), permit IAEA inspectors to return to North Korean sites, and continue the difficult talks toward a lasting resolution was to the United States' ambitions vis-à-vis North Korean threat potentials,[40] the United States was well advised to not pat

---

36 For examples of such Roh pressures, see K. Connie Kang, "Roh Cautions U.S. on North Korea," *Los Angeles Times*, November 13, 2004; Lee Joo-hee and Kim Ji-hyun, "Roh cautions Washington hawks," *Korea Herald*, January 1, 2006; and Andrew Salmon, "Roh Opposes U.S. Regime Change Plans For North," *Washington Times*, January 26, 2006.

37 For examples, see Frank J. Gaffney, Jr., "North Korean Scorecard," *Washington Times*, January 14, 2003; and William Kristol and Robert Kagan, "North Korea Goes South," *The Weekly Standard*, January 20, 2003.

38 Selig S. Harrison in "Roh Balances National Priorities and Alliance with U.S.," *Korea Policy Review*, February 2006 used that term to include Vice President Cheney's Chief of Staff (David Addington), the Deputy National Security Advisor (J.W. Crouch), and the Under Secretary of State for Arms Control (Robert Joseph) because they "orchestrated a campaign to depict North Korea as a 'criminal regime' with which normalized relations are not possible."

39 For an early advocate of that approach, who later moved on to a position at the National Security Council, see Victor Cha, "Hawk Engagement: Bush Policy Toward North Korea," *CSIS Working Paper* (Washington, DC: Center for Strategic and International Studies, 2001).

40 For coverage of that agreement, see Mark Magnier and Barbara Demick, "N. Korea Waives Nuclear Programs," *Los Angeles Times*, September 19, 2005; Gordon Fairclough

itself on the back too visibly. Although U.S. pressures clearly played a major role in inducing Pyongyang to make crucial concessions in exchange for comparatively limited reciprocal offers, it is even more clear that the decisive role in convincing North Korea to change course was played by a combination of China and South Korea.[41] That level of success was temporary, however, due to North Korea's reaction to the larger situation. In part because of the United States' pressures on North Korea the six-party talks were halted for more than a year, during which – on October 9, 2006 – North Korea conducted a successful nuclear weapons test and took great pride in its new status.[42] Shortly after that test, with mounting pressures from all the other five members of the six-party talks, North Korea agreed to return to the dialogue process.[43] Even though the revived talks – December 18-22, 2006 – did not produce any concrete progress, the fact that they were held amounted to de facto progress in terms of getting the process back on track.[44]

Throughout the entire sequence of six-party talks, before and after the interlude, China used its clout as the host-sponsor of the Six-party talks and status as North

---

and Carla Anne Robbins, "North Korea Vows to Give Up Nuclear Programs," *Wall Street Journal*, September 20, 2005; and Donald Kirk and Howard Lafranchi, "North Korea's Agreement to Scrap its Nukes," *Christian Science Monitor*, September 20, 2005. For the text of the Six-party talks joint statement, see "A 6-party Statement: The Deal and Details," (AP), *International Herald Tribune*, September 19, 2005. For some of the nuances in China's role in the negotiations, see Edward Cody, "U.S. Raises Objections to Chinese Proposal In North Korean Nuclear Talks," *Washington Post*, September 18, 2005.

41 For useful coverage of Chinese-South Korean interactions vis-à-vis North Korea, see Chung Jae-ho, "South Korea Between Eagle and Dragon: Perceptual Ambivalence and Strategic Dilemma," *Asian Survey* (September/October 2001); David Shambaugh, "China and the Korean Peninsula: Playing for the Long-Term," *Washington Quarterly*, Spring 2003; Andrew Scobell, *China and North Korea: From Comrades-in-Arms to Allies at Arm's Length* (Carlisle, PA: Strategic Studies Institute, U.S. Army War College, 2004); Lee Guen, "The Rise of China and Korea's China Policy" Kokubun Ryosei and Wang Jisi, eds., *The Rise of China and a Changing East Asian Order*, Tokyo: Japan Center for International Exchange, 2004; and Robert Marquand, "China Changes Game in N. Korea," *Christian Science Monitor*, March 1, 2006.

42 For coverage of the test and reactions to it, see: David E. Sanger, "For U.S., A Strategic Jolt After North Korea's Test," *New York Times*, October 10, 2006; Bennett Ramberg, "North Korea Has the Bomb. Now What?", *Christian Science Monitor*, October 11, 2006; "Special Report North Korea; The Nightmare Comes To Pass," *Economist*, October 14, 2006; Peter Grier, "Pyongyang's Nukes: How Dangerous?" *Christian Science Monitor*, October 19, 2996; and Evan Ramstad, "North Korea Emphasizes Nuclear Pride," *Wall Street Journal*, December 15, 2006.

43 Joseph Kahn and Helene Cooper, "North Korea Will Resume Nuclear Talks," *New York Times*, November 1, 2006; "North Korea's Bomb; Talking Again," *Economist*, November 4, 2006; and Peter Ford, "Nuclear Test Hangs Over N. Korea Talks," *Christian Science Monitor*, December 18, 2006.

44 Joseph Kahn, "Talks End on North Korea's Nuclear Weapons," *New York Times*, December 23, 2006; Mitchell Landsberg, "N. Korea Nuclear Talks End With No Resolution," *Los Angeles Times*, December 23, 2006; Edward Cody, "Nuclear Talks With N. Korea End in Failure," *Washington Post*, December 23, 2006; and Loo Joo-hee, "Nuke Talks End Without Breakthrough," *Korea Herald*, December 23, 2006.

Korea's closest partner, key advisor, and essential benefactor to persuade North Korea to compromise. Equally important, China's recognition of the growing economic and geopolitical bond between the PRC and the ROK, coupled with Beijing's grasp of the importance of the Roh Moo-hyun government's nuanced engagement approach to inter-Korean reconciliation and reunification, put China in an excellent position to be a catalyst for resolving the nuclear controversy in a way that can satisfy the United States, but more importantly be instrumental in accelerating improved inter-Korean relations under China's auspices as a benevolent good neighbor drawing on a long Confucian tradition of Sino-centrism in Eastern Asia. To make this work requires the kind of diplomatic and cultural finesse China subtly displayed in its approach to the Six-party talks by adapting President Roh's engagement policies toward North Korea within the PRC's policies toward both Koreas. This enabled Beijing to make its case to Pyongyang in terms that would benefit the entire Korean nation en route to the shared Korean goal of reunification. China's growing role in Korean affairs is part of the PRC's expanding role in overall Asian regional affairs as well as its globalizing role.[45] North Korea's nuclear bomb test intensified and complicated China's influence.[46] However, China's innovative and flexible approach to the two Koreas—that makes use of the positive aspects of China's traditional historical legacy with Korea and of the negative aspects of both Koreas' modern historical legacy with the United States—put the United States in a more questionable position.

The United States' policies toward Chinese interactions with the two Koreas and attempts to utilize the PRC as an instrument for U.S. efforts to induce the DPRK to alter its nuclear policies and conform to American notions of reform have had a decidedly mixed record.[47] While China has not overtly rejected U.S. initiatives, the way the PRC has cooperated in a limited fashion has sent signals to both Seoul and

---

45 For assessments of both facets of China's expanded roles, see Quansheng Zhao, "China and the Korean Peace Process," Tae-Hwan Kwak and Seung-Ho Joo, eds., *The Korean Peace Process and the Four Powers* (Aldershot, England: Ashgate Publishing, 2003); and Ross Terrill, *The New Chinese Empire; And What It Means for the United States* (New York: Basic Books, 2003).

46 Simon Montlake, "After North Korea's Bomb Test, China Ponders A Problematic Friendship," *Christian Science Monitor*, October 11, 2006; Mark Magnier, "N. Korean Threat Different for China," *Los Angeles Times*, October 13, 2006; Thom Shanker and Joseph Kahn, "U.S. and China Call For North Korea To Rejoin Talks," *New York Times*, October 21, 2006; and David Lague, " Korean Crisis Brings U.S. and China Closer," *International Herald Tribune*, December 19, 2006.

47 In addition to previous citations, see Andrew Scobell, "Crouching Korea, Hidden China—Bush Administration Policy Towards Pyongyang and Beijing, *Asian Survey* (March/April 2002); Douglas Bandow, "Enlisting China: The Battle for Nuclear Free Korea," *National Review*, April 29, 2003; David Sanger and William Broad, "US Asking China to Press North Korea to End Its Nuclear Program," *New York Times*, February 9, 2005; Liping Xia, "The Role of the Korean Factor in China's Foreign Policy," *Korea and World Affairs* (Summer 2005); and Fei-Ling Wang, "A Regional Play of the Global Game: China's Korea Policy and the Sino-American Relationship," *International Journal of Korean Unification Studies*, vol. 14, no. 2, 2005.

Pyongyang that Beijing has its own policy agenda and is not reading from a script devised by Washington.[48] Moreover, as China's political, economic, and security ties with both Koreas grow in ways that signal a more "made in China" than "made in USA" tone for what may be a successful paradigm for improved inter-Korean relations, the PRC's approach to the North Korean nuclear issue gains ground compared to the Bush administration's approach. This trend tends to underscore the ways U.S. policy toward North Korea's nuclear agenda remains troubled.

An Alternative Paradigm

The vision of Bush policy-makers and their advisors is, as explained previously, notably hawkish and not well disposed to work closely with the Roh administration in Seoul. One notable example was an advisor's suggestion that the Bush administration should "work around" President Roh.[49] Fortunately such analyses are confronted by more balanced assessments.[50] So far the Bush administration's harsh pressures exerted on North Korea have not precipitated sufficient internal circumstances to provoke regime change of the sort the most hawkish American opponents of Kim Jong-Il have advocated. Similarly, so far the seriously flawed policies of the Pyongyang government have not been sufficiently inept to lead to any of the collapse scenarios that have been raised over the years by American advocates of a hawkish approach who clearly hoped some such implosion scenario would be accelerated by U.S. pressures. Hence the answer to questions about the overall effectiveness of U.S. policy on North Korean issues must remain negative—i.e., "no." This has had significant impact on the credibility of U.S. policy toward North Korea and on U.S.-DPRK bilateral contacts[51] which, in turn, has had major consequences for U.S.

---

48 This aspect of PRC-U.S. relations was underscored by the United States' geopolitical assessment of China in the most recent *Quadrennial Defense Review Report* (U.S. Department of Defense, February 6, 2006) in which China's evolving position is described as being the most prominent of several countries at a "strategic crossroads" (see pp. 27-32 of the *QDR*). Significantly in a PRC-sponsored section of the *Washington Post*, the Vice Director of the Institute of American Studies at the Chinese Institute of Contemporary International Relations, Yuan Ping, rebutted the QDR's description by reversing the logic in an article entitled "It's US that stands at strategic crossroads," *Washington Post Weekly, Reports from China*, February 27, 2006.

49 Nicholas Eberstadt, "Tear Down This Tyranny," *The Weekly Standard*, November 29, 2004. For background on the Bush administration's troubled approach to North Korea, see Sebastian Harnisch, "U.S.-North Korean Relations Under the Bush Administration—From 'Slow Go' to 'No Go,'" *Asian Survey* (November/December, 2002).

50 For example, see Roland Bleiker, "A Rogue Is a Rogue Is a Rogue: U.S. Foreign Policy and the Korean Nuclear Crisis," *International Affairs*, vol. 79, no. 4 (2003). For a broader assessment of that U.S. foreign policy theme, see Clyde Prestowitz, *Rogue Nation; American Unilateralism and the Failure of Good Intentions* (New York: Basic Books, 2003).

51 The questionable quality and results of such bilateral contacts were ironically symbolized by the awkwardly worded title of an article on U.S.-DPRK discussions on the issue of illegal currency: Annie I. Bang, "N.K., U.S. To Hold Counterfeit Talks," *Korea Herald*, February 25, 2006.

relations with South Korea, China, and Japan. In this context it may be time for Americans and their government to consider the possible benefits of another "no" option for U.S. policy toward North Korea.

At the same time as U.S. policy regarding its nuclear and other problems with North Korea has been so troubled, South Korea's engagement initiatives toward the northern half of the long divided Korean national peninsula under the guidance of President Roh Moo-hyun have been comparatively successful in getting Pyongyang's attention, nudging China to be supportive of the inter-Korean dialogue process, and setting the stage for further progress via the Six-party talks.[52] The approach of the United States' South Korean ally toward North Korea via the Roh policy has been much more successful than the Bush policy has been, notwithstanding Roh's domestic political problems. Washington might well consider adapting a version of the Roh policy toward inter-Korean affairs as U.S. policy toward possible Korean reconciliation and reunification. Bearing in mind how Koreans phonetically pronounce the name Roh—i.e., 'no'—a strong case can be made that the United States needs a new "Roh" policy of its own for coping with North Korea.[53]

In order to make this case, it is necessary to examine how and why U.S. policy is at variance with the policies of the two ROK administrations that have been in office concurrently with President Bush—the Kim Dae-Jung and Roh Moo-hyun administrations. South Korea's Roh policy builds upon the precedents established in Seoul's previous administration under Kim Dae-Jung. While both of these governments' electoral track records have greatly strengthened South Korean democracy compared to the more authoritarian predecessors in Seoul, neither's pro-democracy stance has been sufficient to gain strong rapport with the Bush administration, despite its emphasis on spreading democracy worldwide. President Kim's innovative development of his "Sunshine Policy" of flexible ROK engagement with the DPRK,[54] use of inter-Korean summitry, and willingness to break the client state mold of U.S.-ROK relations en route to cultivating improved peninsular relations gained major recognition in the form of his winning the Nobel Peace Prize. Via these policies President Kim fundamentally changed the tone of South Korea's approach

---

52 For an assessment of those successes by a former ROK Unification Minister, see Park Jae-kyu, "North Korea Since 2000 and Prospects for Inter-Korean Relations," *Korea Policy Review*, February 2006.

53 The author has advocated variations on that theme in previous articles "A Korean Solution to the United States' Korean Problems," *The Journal of East Asian Affairs* (Fall/Winter 2003); "U.S.-North Korean Relations: Foreign Policy Dilemmas," *North Korean Review* (Fall 2005); and "Transforming U.S. Policy Toward Korean Unification: Creating a New U.S.-U.K. Special Relationship," *Korea and World Affairs* (Summer 2005).

54 For background on that policy, see Moon Chung-in and David I. Steinberg, eds, *Kim Dae-Jung Government and Sunshine Policy: Promises and Challenges* (Washington, DC and Seoul: Georgetown University Press and Yonsei University Press, 1999); Norman D. Levin and Yong-Sup Han, *Sunshine in Korea; The South Korean Debate over Policies Toward North Korea* (Santa Monica: RAND, 2002); and L. Gordon Flake, "Sunshine or Moonshine?: Inter-Korean Relations and their Impact upon the U.S.-DPRK Conundrum," in Larry A. Niksch, et al, *Implementing the Six-Party Joint Statement and the Korean Peninsula* (Seoul: Korea Institute for National Unification, 2005).

to North Korea and established the foundation upon which the Roh administration has created a still more flexible set of approaches within the successor "policy of peace and prosperity." These two ROK governments' *very* liberal approach to world affairs—especially when applied to inter-Korean affairs, North Korea's diplomatic brinkmanship strategy, and acceptance of the impact China's growing global clout is having upon Asia—have not bolstered the image of either Kim or Roh in the eyes of the Bush administration. Instead the United States exerted its own forms of pressure upon the PRC to encourage China to be a positive intermediary in the nuclear issue. As much as China is content to let Americans believe what they want to believe about why Beijing helps Washington persuade the Kim Jong-Il government to concede on the nuclear issue, the representatives of China and the two Koreas can sense a different dynamic at work in which Korean responses to both China's stature and its interests in Korea matter more. In particular, President Roh's advocacy of the ROK as a "balancer" in the regional system that reflects the ROK's closer bonds with the PRC and unease about the objectives of U.S.-Japan policies in influencing China's approach to Korea[55] has not been well received in Washington but it has been in Beijing and Pyongyang. Therefore the notion of adapting a Roh policy paradigm within U.S. policy toward North Korea will not be an easy sell even though South Korea before, during, and after this period of advocating a more independent ROK geopolitical role continued to proclaim the importance of the U.S.-ROK alliance and received support from key U.S. officials.[56] That evolving ROK role is compatible with the growing U.S. emphasis on American "strategic flexibility" worldwide which will alter the U.S. military role in Korea.[57] This reinforces how South Korea's balancer role can blend with an evolving U.S. geopolitical paradigm and changes in the U.S.-ROK alliance.[58]

Nonetheless, U.S. adaptation of the Roh paradigm is a worthwhile approach. If the United States were to create an agenda to deal overtly with North Korea as

---

55 For coverage of that approach and domestic South Korean reactions to it, see Unattributed, "Roh Tells U.S. To Stay Out Of Regional Affairs," (UPI), Washington Times, March 11, 2005; Burt Herman, "S. Korea to Play Neutral Role in Asia," *www.newsday.com* (April 10, 2005); and Lee Joo-hee, "Majority of public backs Korea 'balancer' role," *Korea Herald*, April 11, 2005.

56 For examples of those positions, see Shim Jae-yu, "Roh Emphasizes Alliance With US," *Korea Times*, February 26, 2005; Jung Sung-ki, "Seoul-Washington Alliance Remains Solid: USFK Chief," *Korea Times*, April 23, 2005; Gordon Fairclough, "South Korea, U.S. Seek To Mend Their Strained Ties," *Wall Street Journal*, June 7, 2005; Thom Shanker, "Rumsfeld And South Korea Defense Chief Agree To Keep Status Quo," *New York Times*, October 22, 2005; and Jung Sung-ki, "Seoul Agrees On US Troop Flexibility," *Korea Times*, January 21, 2006.

57 Jin Dae-woong, "'Strategic Flexibility' New Credo For USFK," *Korea Herald*, January 21, 2006.

58 For coverage of that evolution, see unattributed, "Roh Stresses S. Korea-U.S. Alliance," *Korea Times*, May 2, 2006; Park Song-wu, "Seoul To Get Back Wartime Command In 5 To 6 Years," *Korea Times*, June 5, 2006; Jung Sung-ki, "S. Korea, U.S. Discuss Future Alliance," *Korea Times*, July 14, 2006; and Jung Sung-ki, "Washington Plans To Transfer Wartime Command By 2009," *Korea Times*, July 20, 2006.

part of the entire Korean nation, in recognition of the ongoing efforts of both the ROK and the DPRK to reconcile their differences, foster peaceful co-existence, and create the foundation upon which a reunited Korean nation state can be built via creative engagement by both sides, the United States would be far better positioned to resolve its diverse differences with North Korea in cooperation with South Korea.[59] Functionally the United States could bolster its institutional and bureaucratic ability to cooperate with the ROK by improving its means for systemic and intellectual dialogue between South Korean and American policy planners. This would be based on the creation of American counterparts to organizations that exist in the ROK government, think tanks, and universities which focus on inter-Korean issues.[60] However, to truly cooperate with South Koreans vis-à-vis North Korea, the United States needs to reassess the ways U.S. policy is not perceived in North Korea as positively as ROK policy is. The best way to do that would be to appreciate the merits of the Roh policy's priorities regarding encouraging North Korean systemic transformation economically that will motivate Pyongyang to be more accommodating with economic partners in South Korea, China, and other countries.[61] The more the DPRK will emulate the state-guided socioeconomic models of the ROK and the PRC, the more likely this will positively influence North Korea politically and strategically.[62]

In order to encourage such North Korean emulation of PRC and ROK reforms via U.S. policy toward the two Koreas, and toward enhancing the prospects for a single Korean nation state at some point in the future, it would be helpful if the United States were to be even more creative in adjusting the U.S.-ROK alliance to evolving circumstances in ways that would appeal to the government in Seoul in a manner that does not disturb ROK-DPRK relations.[63] For the United States to be creative regarding such security issues with its South Korean ally the United States

---

59 For a cautiously critical assessment of what the United States has and has not done to cooperate with South Korea vis-à-vis North Korea, see Young-Ho Park, "Building a Solid Partnership: The ROK-U.S. Policy Coordination on North Korea," Larry A. Niksch, et al, *Implementing the Six-Party Joint Statement and the Korean Peninsula* (Seoul: Korea Institute for National Unification, 2005).

60 The author addressed that issue in greater detail in his "U.S. Policy Planning Toward Korean Unification: A New Approach," *International Journal of Korean Unification Studies* (Summer 2005).

61 For examples of official statements of that policy, see The full text of President Roh Moo-hyun's Liberation Day Address, "Let's Open a New Era of National Unity," *Korea Policy Review* (September 2005); and *Peace and Prosperity: White Paper on Korean Unification 2005* (Seoul: ROK Ministry of Unification, 2005).

62 For a useful example of North Korea's admiration of China as a model, see Kim Jong-Il's speech at the Great Hall of the People during his January 2006 tour of China in which he said, "The progress made in the southern part of China which has undergone a rapid change and the stirring reality of China, in particular, deeply impressed us" and "Our visit to the southern part of China convinced us once again that China has a rosier future thanks to the correct line and policies advanced by the Communist Party of China." in "Speech of by (sic) Kim Jong Il at Banquet," January 18, 2006, online at <www.kcna.co.jp>.

63 For a useful example of what Seoul might want the alliance to become, see Park Kun-young, "A New U.S.-ROK Alliance: A Nine-Point Recommendation for a Reflective and

should adapt the Roh "peace and prosperity" engagement paradigm to U.S. policy and minimize the obstructionist "no" qualities surrounding the Bush administration's approach toward North Korea. U.S. policy makers should learn as many lessons as possible from the ways the ROK attempts to help North Korea improve its economy and its human skills necessary for a more productive economy.[64] The United States can, and should, develop bilateral socioeconomic ties with North Korea designed to provide the DPRK with incentives to pursue reforms, become a participant in regional international trade, and aspire to becoming a player in the forms of globalized trade that most countries rely upon. Such bilateral connections with North Korea should be perceived by American policy makers as a logical result of South Korea's *Nordpolitik* approach to inducing North Korea to develop relationships comparable to South Korea's international connections, bearing in mind that this approach was inspired by West Germany's successful *Ostpolitik* overtures toward East Germany[65] and ultimately set the stage for Kim Dae-Jung's Sunshine Policy overtures toward the DPRK.

Beyond the Nuclear Issue

By halting a negative "no" policy and adapting a proactive "Roh" policy to U.S. purposes the United States would effectively be applying what theorists characterize

---

Mature Partnership," James M. Lister, ed., *Challenges Posed by the DPRK for the Alliance and the Region* (Washington, DC: The Korea Economic Institute, 2005).

64 For examples of the kind of ROK policy parameters vis-à-vis North Korea from which U.S. officials could learn useful lessons, see Kim Young-yoon and Choi Soo-young, *Understanding North Korea's Economic Reforms* (Seoul: Center for the North Korean Economy, Korea Institute for National Unification, April 2005), Hoon Park, "Strategies for Inter-Korean Economic Cooperation: A Managerial Approach," *North Korea Review* (Fall 2005); Choo Won Suh, "Inter-Korean Cooperative Efforts to Normalize North Korea's Industry," Korea Development Institute's *KDI Review of the North Korean Economy* (August 2005) in *Korea Focus* (November-December 2005); Kyuryoon Kim, *Energy Cooperation with North Korea: Issues and Suggestions* (Seoul: Korea Institute for National Unification, 2005); and Kwon Young-kyong, "An Analysis of Trends in North Korea's Economic Reforms in Recent Years," *Vantage Point* (February 2006). As the United States evaluates what lessons can be learned, U.S. officials also should pay attention to why some South Koreans are optimistic and some are less optimistic about the prospects for ROK-DPRK economic cooperation. For an example of an optimistic analysis of the situation, see Kang-Taeg Lim and Sung-Hoon Lim, *Strategies for Development of a North Korean Special Economic Zone through Attracting Foreign Investments* (Seoul: Korea Institute for National Unification, 2005). For an example of a less optimistic analysis, see: Young-Yoon Kim, *Evaluation of South-North Economic Cooperation and Task for Success* (Seoul: Korea Institute for National Unification, 2005).

65 For background on that policy and its context, see Lee Seo-hang, ed., *Evolving Multilateral Security Regime in Northeast Asia* (Seoul: Institute of Foreign Affairs and National Security, 1994).

as "soft power."[66] This use of non-military forms of national power relies upon multiple states' ability to appreciate how their economic, political, and cultural interdependence can cause them to want to avoid hostile acts toward each other. The two Koreas' neighboring state of Japan has been using a form of such strategic harmony building for decades since the mid-Cold War, called *sogo anzen hosho* (comprehensive security).[67] While some observers of Japan's strategic evolution toward a "normal country" status under the leadership of post-Cold War Prime Ministers Koizumi Junichiro and Abe Shinzo suspect it is edging toward reviving military hard power, the roots of Japan's soft power comprehensive security[68] remain very well grounded.[69] Although South Korea was initially reluctant to acknowledge that the ROK was experimenting with that strategic paradigm,[70] over time South Korea's internationalized foreign policy and globalized trade policy made use of it. More important for present purposes, the Sunshine Policy and its "peace and prosperity" successor under President Roh have applied that engagement model to inter-Korean affairs to encourage more interdependence and tension-reduction. The more the United States learns to appreciate the salience of that paradigm the better positioned the United States will be to learn from its South Korean ally about how to effectively cope with North Korea in ways that will be productive for both Koreas, their neighbors in Asia, and—most important for U.S. policy makers—for the United States in its role in the international community. Along those lines, the United States may also consider going beyond the Roh paradigm to explore even more thorough-going forms of multilateral engagement intended to very positively transform North Korea's role in world affairs by encouraging various other countries affiliated with the United States to interact positively with the DPRK.[71] This option may become more salient as Americans recognize the difficulty of resolving the North Korean

---

66 For a prominent example, see Joseph S. Nye, Jr., *Soft Power: The Means to Success in World Politics* (New York: Public Affairs, 2004).

67 Although that doctrine had its roots in the 1960s and 1970s, it was officially launched in Japan by the Ohira administration in 1980. For useful background on the doctrine, see: Robert W. Barnett, *Beyond War: Japan's Concept of Comprehensive National Security* (Washington: Pergamon-Brassey, 1984).

68 For insights into that situation from a prominent theorist who also helped shape U.S.-Japan post-Cold War security ties via the "Nye Initiative," see Joseph S. Nye, "U.S. Power and Strategy After Iraq," *Foreign Affairs*, July 1, 2003 (Op-ed); and Joseph S. Nye, "Asia's Hardening Soft Power," *Taipei Times*, November 17, 2005.

69 In addition to numerous other authors who have analyzed Japan's defense policy's comprehensive balance, the author assessed that issue in *U.S.-Japan Strategic Reciprocity*, Stanford: Hoover Institution Press, 1985 [its Japanese translation: *Bei-nichi keizai boei rinku ron* [U.S.-Japan Economic Defense Link Theory], Tokyo: Jichosha, 1990]; and *Homeland Security: Learning From Japan*, Oakland: The Independent Institute, 2005.

70 The author explored that transition in its formative stage in "Korean Security: Is Japan's 'Comprehensive Security' Model a Viable Alternative?" in Doug Bandow and Ted Galen Carpenter, eds., *The U.S.-South Korean Alliance; Time for a Change* (New Brunswick: Transaction Publishers, 1992).

71 For a survey of such an approach's potential benefits, see Sherry Gray, *Future Multilateral Economic Cooperation With the Democratic People's Republic of Korea* (Muscatine, IA: The Stanley Foundation, June 15-17, 2005).

nuclear crisis largely on U.S. terms. A strong case can be made that the United States should seek resolution of the nuclear problem as part of a broader package of tension-reduction efforts designed to be a catalyst for inter-Korean reconciliation and reunification. Those efforts, in turn, clearly would benefit from an array of international players interacting with North Korea in ways that provide the DPRK leaders with incentives to cooperate on a broad range of issues—obviously including its nuclear agenda.

Yet another example of how the United States may, and probably should, go beyond the Roh paradigm would be preparing for a literal end to the Korean War by going beyond the existing truce arrangements and creating a formal peace treaty between the DPRK and the U.S.-ROK allies. Although this issue has existed ever since the Korean War truce was signed on July 27, 1953, with South Korea abstaining,[72] the fact that it has existed for more than half a century without resolution is a profound example of something which does not receive enough attention by U.S. policy makers. While there has been some scholarly attention paid to the pros and cons of creating a Korean War peace treaty,[73] such analyses have not spawned anything concrete in terms of U.S. policy toward North Korea bilaterally or in tandem with South Korea. This does not surprise most American specialists in Korean affairs who have grown accustomed to the status quo shaped by the truce's legacy.[74] However, given the massive amount of attention the United States pays to many other "post-conflict" and "nation-building" agendas around the world,[75] it *should* be surprising that the United States does not actively plan for either of these agendas with regard

72 The sources cited in note #3 deal with that issue in detail.

73 The most prominent contemporary example is Chapter 13, "Ending the Korean War" in Selig Harrison, *Korean Endgame*, op. cit.. Both sides in Korea periodically advocate creating a peace treaty to formally end the war and move beyond it. For relevant examples, see "N. Korea calls for US peace treaty," *BBC News*, July 22, 2005 online at <www.newsvote.bbc.co.uk>; Ryu Jin, "NK, US to Discuss Peace Treaty," *Korea Times*, August 8, 2005; and unattributed, "South Korea's Roh wants Korean peace treaty talks," *China Daily*, January 25, 2006 online at <www.chinadaily.com.cn>. For very differing views about what the United States should do about a peace treaty with North Korea, see Chalmers Johnson, "Does The US Really Want Peace in Korea?" *Los Angeles Times*, September 8, 2000; and Larry M. Wortzel, "Why North Korea Should Sign A Peace Treaty with The U.N., Not The U.S.," *Executive Memorandum*, Heritage Foundation, May 17, 2001.

74 The author addressed the issue's broader parameters in his *Toward Normalizing U.S.-Korea Relations: In Due Course?* (Boulder: Lynne Rienner Publishers, 2002); and its expanded Korean translation *Hanmi kwangae ui sae jipyung* [New Horizons of U.S.-Korean Relations] (Seoul: Ingansarang Publishers, 2003).

75 For prominent contemporary examples of analyses of this issue, and its relevance for international policy, see James Jay Carafano, *Post-Conflict and Culture: Changing America's Military for 21st Century Missions* (Washington, DC: Heritage Foundation, 2003); Robert C.C. Orr, *Winning the Peace: An American Strategy for Post-Conflict Reconstruction* (Washington, DC: Center for Strategic & International Studies, 2004); Samuel Berger and Brent Scowcroft, *In the Wake of War: Improving U.S. Post-Conflict Capabilities* (New York: Council on Foreign Relations Press, 2005); and Sultan Barakat and Gareth Wardell, eds., *After the Conflict: Reconstruction and Redevelopment in the Aftermath of War* (London: I.B. Tauris & Company, 2005).

to formally ending the Korean War and coping with the "postwar" circumstances that shall exist after a Korean War Peace Treaty is signed. The essence of such U.S. planning for a "post-conflict" scenario on the Korean peninsula would entail far greater U.S. attention to the spectrum of issues inherent in inter-Korean reconciliation and reunification. Such issues constrain both Koreas' policies toward each other and toward all the external non-Korean players' current efforts to resolve the contentious nuclear issue and its impact on short-term and long-term inter-Korean relations. In turn, this context decisively shapes the Roh policies from which the United States should learn.

If Washington would adapt the Roh paradigm within U.S. policy by making Korean unification a much higher priority within U.S. policy toward both Koreas than it has been in the past and is today, it too could build mutual confidence between the United States and North Korea. To do this American leaders would have to stop perceiving such engagement overtures as appeasement and tone down preconditions linked to verifiable nuclear concessions by Pyongyang. This does not mean the United States should "appease" the DPRK or certainly not abandon its nuclear non-proliferation goals that are making progress. Instead, it should learn to appreciate how adapting the Kim-Roh inter-Korean agenda to U.S. policy can instill innovative flexibility within U.S. policy that will enable it to generate incentives in North Korea to reciprocally engage with the United States in ways that will facilitate solid U.S. support for inter-Korean reconciliation and reunification. This would amount to an innovative variation of the "strategic flexibility" paradigm the United States uses within an evolving U.S.-ROK security relationship. A very significant by-product of the mutual U.S.-DPRK flexibility would be creating incentives on both sides—with the support of the ROK and the PRC—for genuine resolution of the contentious nuclear issue as the talks continue to evolve. The United States can indeed get what it wants out of North Korea if it it can learn to be a catalyst for putting into North Korea the kinds of ingredients that will generate the sort of growing multilateral rapport which results from the Roh policy. So, the United States should shed its negative 'no' policy in favor of adapting a positive Roh policy that will better serve both U.S. national interests and the long term national interests of the entire Korean nation. As it does so, the United States also should prepare for ways in which the successor(s) to the Bush administration—Republican or Democratic—can cooperate creatively with the successor(s) to the Roh administration in a manner which will continue to serve both U.S. and Korean national interests.

Chapter 4

U.S. Policy Toward North Korea Under George W. Bush: A Critical Appraisal

Curtis H. Martin

This study describes and evaluates, from the perspective of pertinent system, state, and individual level theory, the unfolding of the Bush administration's strategy for addressing the North Korea nuclear issue up to the February 2007 Beijing agreement on first steps toward denuclearization of the Korean peninsula. It will explore whether that policy has been "effective harmony of different goals," as the administration has claimed, or something more akin to "ostrich engagement," "drift," "mixed signals," "hostile neglect"—or more recently—"Clintonian appeasement" as claimed by critics.[1] At its inception, the administration's policy was grounded in a strong—though by no means universally held—preference for "asphyxiation" rather than "oxygen," for pressure and isolation rather than for rewards.[2] However, this default policy was constrained at almost every turn by system and state level realities that drove the United States to eschew forceful unilateral action and to pursue its objectives through multilateral regional diplomacy. And yet throughout

1 Michael Green, former director of Asian affairs on the National Security Council, cited in Paul Kerr, "North Korea, U.S. Talks Inch Forward," *Arms Control Today* (April 2005). For critical characterizations, see Wade L. Huntley, "Ostrich Engagement: The Bush Administration and the North Korea Nuclear Crisis," Twelfth Annual CANCAPS Conference, December 3-5, 2004 online at <http://www.iir.ubc.ca/cancaps/conf2004/huntley.pdr>; Huntley, "Fiddling While Pyongyang Reprocesses: Bush Administration Folly and the Emergence of Nuclear North Korea," Nautilus Institute, *Policy Forum Online 05-02A*, January 6, 2005 online at <http://www.nautilus.org/fora/security/0502A_Huntley.html>; C. Kenneth Quinones, remarks, "U.S.-North Korean Relations," Paul H. Nitze School of Advanced International Studies, November 2, 2006 lexis-nexis database; Robert Hathaway, "Supping with the Devil" [book review], *World Policy Journal* (Winter 2003/2004), p. 84; Don Oberdorfer, "Does North Korea Have the Bomb?" Johns Hopkins School of Advanced International Studies, *Saisphere* (winter 2004) online at <http://www.sais- jhu.edu/pubaffairs/publications/saisphere/winter04/Oberdorfer.html>. John Feffer, "Caught in the Muddle—Round Two of Bush vs. North Korea," Nautilus Institute, Policy Forum Online 05-15A, February 16, 2005 online at <http://www.nautilus.org/fora/security/0515A_Feffer.html>; Nicholas Eberstadt, "A Real and Present Danger," American Enterprise Online, 2005 online at <http://www.taemag.com/issues/articleID.19258/article_detail.asp>; Bill Gertz, "Bolton Hits Agreement as 'Bad Signal' to Iran," *Washington Times*, February 14, 2007.

2 Franklin Lavin, "Asphyxiation or Oxygen? The Sanctions Dilemma," *Foreign Policy*, 104 (Fall, 1996), pp. 139-53.

the years of six-party talks, the strong preference of core decision makers for isolation continued—pending a more propitious environment to pursue a forward leaning policy—to rein in U.S. negotiators. As a result, U.S. policy often appeared to exhibit characteristics of "temporary appeasement," "hawk engagement" and "malign neglect" in which negotiations assume a primarily tactical role rather than a wholehearted effort to engage the DPRK. The apparent success of financial sanctions, coupled with the international shock wave that followed North Korea's missile and nuclear tests, might have been expected to give administration hawks the long-awaited opportunity to pursue their default policy. Finally legitimized by successive United Nations Security Council resolutions, their quest for tough new sanctions went into high gear. Despite success in broadening sanctions, however, U.S. policy remained as constrained as ever by the grave deterioration of the United States' position in the Middle East and the unanticipated shift of power in Congress. These constraints may in part explain the dramatic about-face in the administration's position that led ultimately to the February 13, 2007 agreement to offer the DPRK an "early harvest" in exchange for initial steps toward denuclearization.

Critics have argued that the Bush administration's passivity, distraction, and confusion allowed the DPRK to pursue their goal of a nuclear weapons capability far more successfully than it had prior to January 20, 2001.[3] Setting aside the question of whether the Bush policy will or will not eventually succeed in achieving its denuclearization objectives, it is the contention here that U.S. policy has been more coherent than critics have allowed. Apparent inconsistencies have to a degree reflected different tacks on a deliberate course, responses to shifting winds in an ever-changing diplomatic and security seascape. It would have been surprising, for example, if U.S. policy toward North Korea had not been influenced by 9/11, the prolonged insurgency in Iraq, China's or South Korea's resistance to sanctions, or revelations about Iran's nuclear program. The need to take into account complicated issue linkages as well as domestic and alliance politics force any administration into contradictory words or actions. Neither the policies of George H. W. Bush nor those of Bill Clinton were free of internal divisions and inconsistencies over North Korea.[4] Finally, critics' charges of "no policy" may be a proxy for "not the policy I support." With the diplomatic game still in progress, any conclusions about the efficacy of the current administration must remain tentative.

Seeking Its Own Path

Influenced by the view that both the 1994 Agreed Framework and the South Korean "sunshine" policy had failed and a conviction that "time was on our side" as the DPRK slid relentlessly toward collapse, the administration made clear that it was in

3 Kongdan Oh and Ralph C. Hassig, "North Korea's Nuclear Politics," *Current History*, September 2004, p. 278. Joseph Cirincione and Jon B. Wolfsthal, "No Good Choices: The Implications of a Nuclear North Korea," *Brown Journal of World Affairs*, 12 (1) (Fall 2005), p. 269.

4 Joel S. Wit, Daniel B. Poneman and Robert L. Gallucci, *Going Critical: The First North Korean Nuclear Crisis* (Washington, D.C.: Brookings, 2004), pp. 139-40,

no hurry to resume the Clinton administration's dialogue. It therefore devoted the first six months of 2001 to a policy review. While not directly rejecting the Agreed Framework, critics within the administration believed that it had "rewarded bad behavior" without ensuring the dismantlement of the North's nuclear program. The United States had even become "a major patron and financier of the survival of the North Korean state."[5] The Agreed Framework had failed to secure a full accounting of all nuclear materials as required by the International Atomic Energy Agency (IAEA), and it had been concluded without the formal participation of U.S. allies South Korea and Japan (though in fact they had been extensively consulted and played a central part in bankrolling the promised light-water reactors). The policy review culminated in a proposal for expanding the negotiating agenda with North Korea to include not only implementation of the Agreed Framework, but an end to missile development and exports, and reduction of the DPRK conventional military threat as well. Determined to send a message that the DPRK faced a new day in U.S. diplomacy, the Bush administration was happy to let the DPRK "cool its heels."

The September attacks on the World Trade Center towers and Pentagon bumped North Korea down the administration's agenda. At the same time, they radically reshaped U.S. perceptions of the North Korea issue, locating it in the context of a global war on terror in which stateless organizations, supported by state sponsors, posed the greatest threat to U.S. security. Accordingly, the administration concluded that the transfer of nuclear materials or technology by the DPRK rather than the possession of nuclear weapons per se was the ultimate "red line." By February 2002, President Bush had included the DPRK in his "axis of evil," though he disavowed any intention of invading it and continued to stress a diplomatic solution. The administration floated proposals for a "bold approach" including substantial economic assistance in exchange for enhanced inspections of the North's nuclear facilities, an end to missile sales, a conventional forces pullback, and—the latest additions to the U.S. demands—an end of support for terrorism and improvement in human rights.

Responding to the DPRK Breakout from the Nuclear Nonpoliferation Treaty

Not until the autumn of 2002 did formal North Korean-United States contacts resume. In October, Assistant Secretary of State for East Asian and the Pacific James Kelly met with North Korean officials, not in order to resume negotiations, but to spell out to the North Koreans the new American ground rules and to confront them with the assertion that the U.S. had evidence of a clandestine uranium (HEU) processing

5 Nicholas Eberstadt, "Towards a Peaceful Resolution with North Korea: Crafting a New International Engagement Framework," American Enterprise Institute, March 8, 2004 available online at <www.aei.org//publications/pubID.20062,filter.all/pub_detail.asp>. The administration view did not acknowledge that substantial sticks accompanied the "carrots" in the Agreed Framework, and that these played an important role in gaining significant North Korean concessions. See Curtis H. Martin, "Rewarding North Korea: Theoretical Perspectives on the 1994 Agreed Framework," *Journal of Peace Research*, 39 (1) (January, 2002), pp. 61-62, 64.

program. In rapid succession talks were suspended and the Agreed Framework declared dead by both sides. Within a few more weeks, the DPRK had withdrawn from the Nuclear Nonproliferation Treaty (NPT) and removed fuel rods from the Yongbyon reactor. During March and April, tensions escalated, with the United States mobilizing military forces in and around Korea, and the North declaring that it both possessed nuclear weapons and had the right to transfer the technology.[6]

Following the demise of the Agreed Framework, the Bush policy coalesced around complete, verifiable, and irreversible disarmament (CVID) and a campaign to increase international isolation of the DPRK. In pursuit of these objectives, the United States would adhere to the following principles: (1) refusal to engage in direct bilateral talks, (2) refusal to offer rewards for bad behavior, and (3) withholding of any benefits until after CVID was assured (if not completed), and (4) delaying full diplomatic relations until after the North addressed additional security and human rights concerns.

To achieve the pressure side of its policy, the administration initially settled upon "tailored containment," a strategy that was never fully defined but appears clearly to have reflected the desire to isolate Pyongyang through a combination of cargo inspections, financial and other sanctions validated, it was hoped, by the United Nations.[7] Instead of confronting the DPRK head on, however, the administration was soon compelled by the looming Iraq conflict and allied resistance to choose more indirect means. Even though direct pressure was largely foreclosed, the administration expected that a preventive war against one axis of evil country would teach the DPRK and other potential proliferators a lesson. It also took advantage of increasingly close relations with Japan not only to deliver an implied threat to the DPRK, but also to remind China that failure to disarm the DPRK would confront it with a more dangerous security environment. Military contingency planning was accelerated, and Bush warned that military force could not be ruled out if diplomacy failed. Pursuant to the Illicit Activities Initiative in 2003, the United States created interdepartmental committees to monitor alleged DPRK criminal activities and mobilized relevant agencies to curtail illegal financial transactions.[8] More visibly, the government inaugurated the Proliferation Security Initiative (PSI), a plan to marshal multilateral efforts to cut off transfers of Weapons of Mass Destruction (WMD)-

6 Curtis H. Martin, "The 2002/2003 Dispute over North Korea's Nuclear Weapons Programs: Spiral or Deterrence Model?" International Studies Association annual meeting, Montreal, Canada, March 16-20, 2004. South Korea's vehement opposition to U.S. air strikes may have persuaded the Bush administration of the need to prove to its allies the futility of negotiations before it could enlist their support for tougher policies. Gregory Elich, "Hawk Engagement: A Dangerous Turn in US Plans for North Korea," Center for Research on Globalization, November 30, 2004 online at <http://www.globalresearch.ca/articles/ELI411A.html>.

7 Michael R. Gordon, "U.S. Readies Plan to Raise Pressure on North Korea," *New York Times*, December 29, 2002. For the DPRK reaction, see "KCNA Blasts U.S. 'Tailored Containment' Strategy," KCNA, January 30, 2003 online at <http://www.kcna.co.jp/item/2003/200301/news01/30.htm>.

8 Ruediger Frank, "The Political Economy of Sanctions against North Korea," *Asian Perspectives*, 30 (3) (2006), p. 29.

related materials by sea or by air. As the centerpiece of the tailored containment policy, PSI eventually emerged as a template for broader sanctions during Bush's second term.

In July of 2003, then Undersecretary of State John R. Bolton summed up U.S. policy as a combination of pressure—to be anchored in PSI and Security Council action (despite the failure of the Council to take any action up to that point) on the one hand, and multilateral negotiations on the other. Notwithstanding the above moves, neither the United States, the United Nations, nor other great powers imposed effective physical or economic penalties on the DPRK for its actions. Until 2006, U.S. efforts to gain Security Council condemnation of the DPRK's actions were stalemated. The U.S. military, poised to attack Iraq, was unavailable for either threat or counter-proliferation duty. Robert Litwak attributed the U.S. posture to "a pragmatic pivot" from regime change to "deterrence and reassurance."[9] Conservative commentator Charles Krauthammer had a somewhat different interpretation, writing in 2003 that the administration's policy of "temporary appeasement," was an expedient that would be "immediately withdrawn as soon as the Iraq war [was] over and we can marshal enough strength in the Northern Pacific to credibly threaten military action [against the DPRK]."[10] Lacking short-term options, U.S. officials took great pains to insist that there was no crisis at all. They publicly reiterated that the United States had "no aggressive intent" and would not seek Security Council sanctions. James Kelly held out hope for more explicit security assurances and economic assistance if the North agreed to give up its nuclear programs.[11] The administration announced its readiness in "a multilateral fashion" to pursue its "bold initiative." The DPRK may have been left in some doubt about how seriously the United States and international community took its breakout from the NPT.[12]

Restarting Talks

Acting to implement Bolton's "multilateral negotiations" track, the United States, North Korea, China, Russia, South Korea, and Japan began in August 2003 what have been dubbed the "six-party talks." The new format was intended to resolve many of the problems that the Bush administration associated with the Agreed Framework. It

9 Robert S. Litwak, "Nonproliferation and the Dilemmas of Regime Change," *Survival*, 45, (4) (Winter 2003-04), p. 8.

10 Charles Krauthammer, "Temporary Appeasement for North Korea," *Townhall.com,* May 7, 2003 available online at <http://www.townhall.com/opinion/columns/charleskrauthammer>.

11 Jay Solomon, Charles Hutzler and David S. Cloud, "A Wall Street Journal Report: U.S. Struggles to Ease the North Korean Standoff-Bush's Envoy Suggests Possibility of Energy Aid to End Nuclear Impasse," *Wall Street Journal*, January 13, 2003, p. A9.

12 Nicholas Eberstadt, "A Real and Present Danger," American Enterprise Online, 2005 available on line at <http://www.taemag.com/issues/articleID.19258/article_detail.asp>; Wade L. Huntley, "Goliath's Game: U.S. Policy toward North Korea in Strategic Context," Presented at the workshop "America in Question: Korean Democracy and the Challenge of Non-Proliferation on the Peninsula," Seoul, ROK, May 10-11, 2005. David E. Sanger, "Bush Shifts Focus to Nuclear Sales by North Korea," *New York Times*, May 5, 2003, p. A1.

was consistent with rejection of one-on-one discussions between the United States and the "evil" DPRK. It would redefine the proliferation issue as a regional, not a bilateral issue and hence would make China and the other powers stakeholders in the negotiations and any agreements reached. A core assumption of the new format, with China as host, was that it would enlist North Korea's main patron in the project of increasing pressure on the regime. The United States expected that North Korea would find itself isolated five-against-one at the table. However, instead of the alignment that the United States had hoped for, a "1-2-3" lineup of the participants quickly emerged. South Korea, with its "peace and prosperity" policy joined China and Russia as the "3" that were most favorably disposed toward conciliation. While Japan paired with the United States as one of the "2" who favored more pressure in the mix, it was still more resistant to a military solution that might provoke a North Korean attack on Japanese territory. The result of these tensions was to reduce the early rounds of the talks to what a critic labeled "diplomatic shadow boxing."[13]

Prior to the second session, the United States relented somewhat by agreeing in principle to coordinating security assurances and verifiable steps at dismantlement as well as to allowing its partners to offer economic assistance prior to CVID.[14] At the third round in June 2004, the United States offered more specific rewards to North Korea for making the correct "strategic decision," though it is reported that opposition within the administration cut short that approach.[15] The United States publicly took a more tolerant position of South Korea's engagement policy but continued to insist that the United States would offer no tangible rewards until the North's nuclear program had been completely dismantled.

Bush's Second Term, a Shift of Emphasis, not Strategy

Strengthened by the 2004 elections, the administration had an opportunity to reassess its North Korea policy. Although Republican gains would seem to have strengthened the hand of hawks, a number of observers concluded that Bush was prepared to pursue a very different tack. Some concluded that the new foreign policy team placed greater emphasis on multilateralism, a willingness to downplay military solutions, and more flexibility on the North Korea issue.[16] The *Wall Street Journal* noted "a

13 James Clay Moltz and C. Kenneth Quinones, "Getting Serious about a Multilateral Approach to North Korea," *The Nonproliferation Review* (Spring 2004), p. 136.

14 *North Korea: A Chronology of Events, October 2002-December 2004*, CRS Report for Congress, RL32743

January 24, 2005 online at <http://digital.library.unt.edu/govdocs/crs//data/2005/upl_meta-crs-6252/RL32743-2005Jan24.pdf>.

15 David E. Sanger and Eric Schmitt, "Cheney's Power No Longer Goes Unquestioned" *New York Times*, September 10, 2006, p. A1, 22.

16 Jay Solomon and Neil King, Jr., "Diplomatic Relations: As 'Neocons' Leave, Bush Foreign Policy Takes Softer Line; Ms. Rice Changes Approach to Iran and North Korea; Democracy Still Key Goal; Cheney's Waning Influence?" *Wall Street Journal*, February 6, 2006, p. A1; Litwak, "Non-proliferation and the Dilemmas of Regime Change," loc.cit., p. 8.

return to argument, persuasion and rhetoric."[17] What emerged at first, however, was not a disjuncture but a modification of the existing two-track U.S. policy of negotiation and pressure.[18] Despite the DPRK's public declaration in February that it possessed nuclear weapons, the Bush administration undertook the most intensive effort at negotiation since the start of six-party talks. Simultaneously, however, it moved to a more confrontational posture.

Stepped up Pressure

Blocked from pursuing formal sanctions, U.S. policy increasingly sought alternative means to squeeze the DPRK. It sought to employ PSI-like coalitions of the willing to cut off the North's revenue from counterfeiting, money laundering and drug trafficking. Acting under the Patriot Act and the Iran, North Korea and Syria Nonproliferation Act, the United States imposed a range of financial sanctions on North Korean trading entities, foreign banks and companies that had dealings with North Korea.[19] Under the Patriot Act the United States was able to designate the DPRK a "primary money-laundering concern," a move that severely constricted third party dealings, notably by South Koreans, with the DPRK.[20] The new sanctions could be pursued independently of the United Nations and appeared to provide a credible and effective threat to the North's economy. Despite the continued drag of the Iraq situation, the administration believed that its financial sanctions had greatly strengthened the United States' bargaining position against what it continued to see as a tottering regime, as well as augmenting its leverage with South Korea and China.[21]

Energizing the Negotiating Track

In tandem with stepped up pressure, the administration invested a substantial effort into energizing the six-party talks and cooling, at least for a time, the rhetorical assaults on the Kim regime. Greater U.S. flexibility led to the Joint Statement at

---

17 Eliot A. Cohen, "From Colin to Condoleezza," *Wall Street Journal*, November 17, 2004, p. A16.

18 Charles L. Pritchard, "Six-party Talks Update: False Start or a Case for Optimism," presented at The Changing Korean Peninsula and the Future of East Asia, Center for Northeast Asian Policy Studies, the Brookings Institution Seoul Forum of International Affairs, December 1, 2005.

19 Julia Choi and Karin Lee, "North Korea: Economic Sanctions and U.S. Department of Treasury Actions, 1955-September 2006," The National Committee on North Korea, October 18, 2006 online at <http://www.ncnk.org/ncnk>.

20 "Seoul Fears U.S. Could Brand N. Korea a Financial Pariah," *Chosun Ilbo*, April 4, 2006. Cited in Nautilus Institute, *NAPSNET Daily Report*, April 4, 2006,

21 Under Secretary of the Treasury Stuart Levy reportedly said that financial sanctions were putting 'huge pressure" on the DPRK. Christian Caryl, "Pocketbook Policing," *Newsweek*, April 10-17, 2006.

the fourth session of talks in September 2005.[22] The statement affirmed the goal of "verifiable denuclearization," including a DPRK commitment to abandon *all* nuclear weapons and *existing* nuclear programs (allowing the United States to interpret the agreement as including the alleged uranium enrichment program), a return to NPT and IAEA safeguards, and an agreement to abide by the principles of the UN Charter and recognized norms. Both sides affirmed the North-South Joint Declaration on Denuclearization of the Korean Peninsula. For its part, North Korea gained written affirmation that South Korea was free of nuclear weapons and would remain so, and U.S. reaffirmation that it had no intention to attack, invade or threaten the DPRK with either nuclear or conventional weapons. Also included was a commitment to observe "mutual respect for each other's sovereignty," to "exist peacefully," and to take steps toward full normalization of political and economic relations (though including none of the specific markers along the way that were in the Agreed Framework). Of great importance to the economic well-being of the DPRK was an expression of willingness by the five powers to provide energy assistance, recognition of the DPRK's right to "peaceful uses of nuclear energy" and an agreement to discuss "at an appropriate time" the provision of a LWR. The statement even included a pledge to negotiate a permanent peace regime at an appropriate separate forum. No specific mention was of side issues such as human rights, conventional military deployments, missiles, counterfeiting, or drug trafficking.

Though the joint statement sustained the U.S. commitment to CVID, it appeared to indicate important shifts in the U.S. position. Language agreeing to "coordinated steps ... in a phased manner" stepped back from the U.S. position that there should be no rewards until CVID.[23] The document put in writing previous administration statements that the United States had no intention of attacking North Korea, and went beyond those assurances in its specificity. It shared several points in common with the Joint Statement of June 1993, the Agreed Statement of July 1993, the Agreed Framework of 1994, and the 1998 Clinton statement that the United States "no longer holds hostile intent."[24] South Korea's offer of energy assistance echoed the Agreed Framework's provision to "offset the energy foregone due to the freeze," while Ambassador Hill's statement that the United States was "prepared to address the DPRK's energy needs" was reminiscent of U.S. heavy fuel oil deliveries under the Agreed Framework.[25] The administration even conceded the right, however

---

22 Larry Niksche, *North Korea's Nuclear Weapons Program,* Congressional Research Service, CRS Issue Brief for Congress, IB91141, August 31, 2005.

23 Paul Kerr, "North Korea Talks Achieve Breakthrough," *Arms Control Today*, October 2005 online at <http://armscontrol.org/act/2005_10/OCT-NKBreakthrough.asp>.

24 See Ha Young-sun, "Meeting the North's Demands," Nautilus Institute, *Policy Forum Online 05-83A,* October 13, 2005 online at <http://www.nautilus.org/fora/security/0583Ha.html>.

25 Tae-Hwan Kwak, "The Six-Party Nuclear Talks and the Korean Peninsula Peace Regime Initiative: A Framework for Implementation," in Tae-Hwan Kwak and Seung-Ho Joo, eds., *The United States and the Korean Peninsula in the 21st Century* (Aldershot, England: Ashgate, 2006), p. 25.

hedged, of the DPRK to obtain light water reactors—a provision at the heart of the Agreed Framework.[26]

The purpose of the United States' sudden flexibility is subject to dispute. Some observers saw it as tactical, even cynical. Henry Sokolski, a longstanding critic of the Agreed Framework, explained the shifting U.S. policy this way: "You have to make the diplomatic effort to give them a chance, but really you're out to prove to everybody that these guys are beyond the pale. ... You move to what is being called Plan B, not a hot war but isolating them as best we can."[27] Victor Cha's claim in the summer of 2002 that the administration was pursuing "hawk engagement" as "an instrument for revealing Pyongyang's unreconstructed intentions" and building support for "punitive actions" reflected the same assessment.[28] There is certainly prima facie evidence that the administration was not fully behind diplomacy. Within days of the joint statement, written interpretations of the statement submitted by each of the six parties revealed potentially unbridgeable differences.[29] U.S. officials emphasized that on a key issue—the timetable for DPRK rewards—the United States had not changed its position.[30] Parting company with other participants, the United States stressed that the "appropriate time" for discussion of a light water reactor was after CVID, after the North had returned to full compliance with NPT and IAEA safeguards, and after it had demonstrated "a sustained commitment to cooperation and transparency and has ceased proliferating nuclear technology." The fifth round of talks in November, scheduled to work out details of implementation of the joint statement, adjourned with no further movement toward the goals of the statement.

The developing U.S. hard line on the joint statement was buttressed by developments in the administration's "pressure" track. Even before the end of the September session of the six-party talks, the U.S. Treasury announced its intention to impose financial sanction on Macao's Banco Delta Asia, a move that was the

---

26 Joseph Kahn and David E. Sanger, "U.S.-Korean Deal on Arms Leaves Key Points Open," *New York Times*, September 20, 2005 online at <http://nytimes.com/2005/09/20/international/asia/20korea.html>.

27 Quoted in Howard LaFranchi and Amelia Newcomb, "In North Korea talks, a glimmer of hope," *Christian Science Monitor*, July 25, 2005 online at <http://www.csmonitor.com/2005/0725/p02s02-usfp.html>.

28 Victor D. Cha, "Korea's Place in the Axis," *Foreign Affairs*, 81 (3) (May/June 2002), p. 81, 89. Kwak notes that U.S. acquiescence in the Joint Statement of September 2005 came about because both Bush and Rice "were persuaded that if a confrontation ever occurred with North Korea, they had to be prepared to show that they had made every effort to reach a diplomatic solution." Kwak, "The Six-Party Nuclear Talks," loc. cit, p. 25.

29 Peter Hayes, David von Hippel, Jungmin Kang, Tadahiro Katsuta, Tatsujiro Suzuki, Richard Tanter and Scott Bruce, "Light Water Reactors at the Six-party Talks: The Barrier that Makes the Water Flow," Nautilus Institute, *Policy Forum Online,* 05-78A, September 21, 2005 online at <http://www.nautilus.org/fora/security/0578LWR.html>.

30 Paul Kerr, "North Korea Talks Achieve Breakthrough," *Arms Control Today*, loc. cit. "Interview with Secretary of State Condoleezza Rice by Time Magazine's Editorial Board, September 20, 2005," U.S. Department of State via Lexis-nexis database September 30, 2005; "The Six-party Talks and the North Korean Nuclear Issue: Old Wine in New Bottles," Hearing, U.S. House International Relations Committee, October 6, 2006 via Lexis-nexis database. Pritchard, "Six-party Talks Update: False Start or a Case for Optimism," loc.cit.

opening salvo in a strategy of cutting off the DPRK's foreign sources of revenue and persuading other governments and financial institutions to do likewise.[31] To underscore the U.S. position on the LWR question, and to draw a curtain over any comparisons with the Agreed Framework, Ambassador Hill announced that the last tenuous link to the Clinton agreement, the Korean Energy Development Organization (KEDO), would soon be terminated.[32] The United States also broke with prior policies by agreeing to admit political refugees from the DPRK into the United States. President Bush's revival of the charge that Kim Jong-Il was a "tyrant" and his reassertion that "a nuclear armed North Korea will not be tolerated" suggested that any perceived appeasement of the DPRK was indeed temporary and tactical.[33] The United States and South Korea pressed ahead with annual military exercises that the DPRK always viewed as provocative. There were even suggestions that the administration was developing plans to bring criminal charges against Kim Jong-Il, much as it had done against Manuel Noriega in 1989.[34]

U.S. Policy Following the DPRK Missile and Nuclear Tests

If there were any doubt about the United States' preference for a policy of isolation, one would have expected it to be removed following North Korea's missile launches in July and nuclear test in October 2006. The DPRK's twin affronts achieved what no previous events could: unite Russia, Japan, China and South Korea behind the United States in Security Council condemnation of the DPRK. Suddenly implementation of the third leg of the policy announced by John Bolton three years previously—Security Council action—was activated.[35] The Security Council resolutions represented the first time in five years that the diplomatic advantage, which had steadily shifted in favor of the DPRK, shifted back toward the United States. Though not going as far as the United States wished, Security Council Resolution 1695 (SC Res. 1695) and Security Council Resolution 1718 (SC Res. 1718) strongly condemned North Korea's actions. SC Res. 1695 demanded suspension of missile testing, required member states to prevent all trade with the DPRK in missiles and related materials, and strongly urged the DPRK to return to the six-party talks. SC Res 1718 went

31 The idea of disrupting financial networks appears in the Pentagon's war plan for North Korea, Op Plan 5030, in 2003. Bruce Auster and Kevin Whitelaw, "Upping the Ante for Kim Jong-Il," *U.S. News & World Report*, July 21, 2003 online at <http://www.usnews.com/usnews/news/articles/030721/21korea.htm>.

32 The Korean Energy Development Organization was created to implement the Agreed Framework promise of constructing more "proliferation proof" light water reactors in North Korea to replace the existing graphite reactors.

33 Charles Krauthammer, "Temporary Appeasement for North Korea," *Townhall.com*, March 7, 2003, loc. cit.

34 "U.S. May Seek Criminal Charges against Kim Jong-Il," *Chosun Ilbo*, March 26, 2006, cited in NAPSNET Daily Report, March 27, 2006, Nautilus Institute.

35 James Cotton called the resolution "a watershed in the Korean nuclear crisis." James Cotton, "Missiles and Sanctions: Has a Watershed Been Reached in the Korean Nuclear Crisis?" *Austral Policy Forum 06-27A*, Nautilus Institute at RMIT, August 10, 2006 online at <http://nautilus.rmit.edu.au/forum-reports/0627a-cotton.html>.

even further, enshrining the U.S. demands for abandonment of *all* nuclear weapons and existing nuclear programs "in a complete, verifiable and irreversible manner." Though calling for resumption of talks based on the 2005 Joint Statement, it imposed on all UN members the obligation to prohibit both imports and exports of WMD-related and other military materials as well as luxury goods, to freeze assets capable of supporting WMD or missile programs, and to take cooperative actions (a la PSI) to inspect cargo. The unanimity of the Security Council undercut what had previously been the DPRK's ace in the hole: deadlock at the UN. For the United States, it provided coveted legitimacy for the PSI and its more recent sanctions.

Certainly the Bush administration would have seen North Korea's behavior as a validation of its hard-line positions. Although the ROK, China, and Russia continued to differ with the United States in significant ways over implementation, the DPRK leadership would have been put in some uncertainty about how these countries would act in the future should the DPRK fail to cooperate.[36] In the wake of the UN resolutions, and with the blessings of the new North Korea Nonproliferation Act, the United States pressed other countries, especially China and the ROK, to impose a wide range of sanctions.[37] Because of the harsh DPRK reaction to the new sanctions, there was initially doubt that the six-party talks would reconvene. But when they did in December 2006, the United States was closer to marshalling a five-to-one dynamic against the DPRK than it had ever been.

Significantly, however, the administration did not use the DPRK tests as an excuse to abandon negotiations. Instead, it insisted that it was still pursuing a "dual-track" policy.[38] Presidential statements continued to call for patient diplomacy, and Undersecretary of State Robert Joseph was reported to have said that the administration's tough measures were part of a strategy of "squeeze them but keep the negotiations going" in the belief that more pressure would "reinforce the prospect for success" at the talks.[39] At this point, the administration made a strategic decision to offer specific, costly concessions to the DPRK in order to break the deadlock.[40]

---

36 At the same time, the ROK surprised the DPRK by for the first time imposing sanctions of its own, in the form of suspending some humanitarian assistance until the North returned to talks. Anthony Faiola, "S. Korea Suspends Food Aid to North," *Washington Post*, July 14, 2006, p. A18.

37 Julia Choi and Karin Lee, "North Korea: Economic Sanctions and U.S. Department of Treasury Actions, 1955-September 2006," loc. cit.; Josh Meyer, "Squeeze on North Korea's Money Supply Yields Results," *Los Angeles Times,* November 2, 2006 online at <http://www.latimes.com/news/nationworld/world/la-fg-norkor2nov02,1,7740617.html>.

38 Undersecretary of State Nicholas Burns told Congress in November, 2006, that U.S. policy was a "dual track policy" of enforcing sanctions, pursuing "a very tough counterproliferation regime, missile defense on the one hand and resuming negotiations through the six-party talks on the other. "North Korea's Nuclear Test: Next Steps," Hearing, U.S. Congress, House International Relations Committee, November 15, 2006.

39 Leon V. Sigal, "An Instinct for the Capillaries," Nautilus Institute, *Policy Forum Online, 06-36* online at <http://www.nautilus.org/fora/security/0636Sigal.html>.

40 Helene Cooper and David E. Sanger, "U.S. Offers North Korea Aid for Dropping Nuclear Plans," *New York Times*, December 5, 2006 online at <http://www.nytimes.com/2006/12/06/world/asia/06korea.html>.

Even when the DPRK agreed to return to the six-party talks in December 2006, it was unclear whether it intended to use the talks to buy more time, or to engage in serious hard bargaining based on the new US proposals. When that session of the talks ended, many feared that it was the former.

However, a bilateral meeting between United States and DRPK negotiators in Berlin in January 2007—itself a significant relaxation of U.S. opposition to bilateral talks outside the six-party format—paved the way for resumption of the Beijing talks and a diplomatic breakthrough. Rather than the "grand bargain" or "bold initiative" that some had urged over the years, the Beijing agreement focused narrowly on small, initial steps toward the disabling and eventual dismantlement of the DPRK's nuclear programs, as provided in the 2005 Joint Statement. All sides agreed to a set of initial actions within a sixty-day framework that would start the parties down the road to a denuclearized Korea. The important concessions from the United States included expressed willingness to contribute to some front-loaded energy assistance, a pledge to "begin the process of removing the designation of the DPRK as a state-sponsor of terrorism," to "advance the process" of removing economic sanctions, and to initiate bilateral talks on other outstanding issues including diplomatic relations. By scheduling a ministerial conference to review progress, the United States implicitly agreed to the highest level diplomatic contacts since the end of the Clinton administration. The agreement marked an important turning point in U.S. policy after six years of refusal to "front load" rewards to the DPRK. Though not ruled out for the future, the "default" policy of isolation appeared to be postponed once again.

Analysis

The Bush policy toward North Korea on the nuclear issue can be interpreted from the perspective of Waltz's three levels of analysis in international relations theory.[41] U.S. policy has reflected a complex regional and international context, a specific political and policymaking context within the United States, and the influence of strong-willed senior policy actors.

System Level

System level analysis focuses on supra-national environmental influences on state behavior such as the identity of state and non-state actors, distribution of power in the system, and the nature of political and economic interactions among actors. System polarity (or perceptions of it) has played a significant part in shaping post-Cold War United States policy, including policy toward the DPRK. The Bush administration's acceptance of a unipolar paradigm can explain many, though not all, of its policy

41 Kenneth N. Waltz, *Man, the State, and War* (New York: Colombia University Press, 1954). These familiar, "first," "second," and "third" images are admittedly not the only "levels" proposed by theorists, but they will serve as well as others for present purposes. Robert Jervis, *Perception and Misperception in International Politics* (Princeton, NJ: Princeton University Press, 1976), p. 15.

choices. To many observers, the United States seemed to confirm the maxim that as a state's power expands, so does its perception of its interests and goals.[42] The sheer size and versatility of U.S. military power permitted U.S. leaders to believe that quick victory in Iraq would "shock and awe" other members of the axis of evil into calculating that the United States would "go down the list" once Iraq had been disposed of.[43] With respect to Asia, Van Ness has written that, "[T]he fact of overwhelming American power plus the acquiescence of the other major powers in East Asia to the U.S. leadership role confirmed the reality of the American hegemonic position."[44] U.S. responsibilities as the global unipole place any individual policy challenge in a global perspective, and since 9/11 the United States has viewed its Korea policy increasingly through the lens of its anti-terrorism policy. The Bush doctrine and supporting documents amount to a declaration that from its hegemonial position, the United States can achieve "absolute security" through unmatchable levels of defense spending and application of the revolution in military affairs (RMA) in selective resort to preemptive war. The neoconservative goal of democratic transformation was consonant with theories of hegemonic stability.[45]

According to Robert Jervis, the opportunities for the United States to exploit its hegemony have been enhanced by the formation of a large security community among most of the large industrial powers.[46] The United States has had reason to be confident, despite the estrangement occasioned by the Iraq war, that though members of this community and other great powers may "seek to harness and restrain American power, they will not attempt to displace it."[47] The U.S. decision to pursue multilateral rather than unilateral policies to deal with the North Korean threat may not have represented an acknowledgement of "multipolarity" or restraint. Instead it may have been a "narrowly instrumental" policy designed more to press the other parties to conform to U.S. designs than to engage in genuine consultative diplomacy.[48] China's and South Korea's gradual movement toward the U.S. position on sanctions may

---

42 Robert Jervis, *American Foreign Policy in a New Era*, loc. cit., p. 94.

43 Ibid., p. 94.

44 Peter Van Ness, "Why the Six-party Talks Should Succeed," Nautilus Institute, *Policy Forum Online* 05-62A, July 27, 2005 online at <http://www.nautilus.org/fora/security/0562Van Ness..html>.

45 Jervis notes the specific endorsement of the concept in the Nuclear Posture Review and its antecedent draft defense guidance document of 1992. Robert Jervis, *American Foreign Policy in a New Era,* loc. cit., p. 90. On hegemonial stability, see A.F.K. Organski, *World Politics* (New York: Knopf, 1968); Paul M. Kennedy, *The Rise and Fall of the Great Powers: Economic Change and Military Conflict from 1500 to 2000* (New York: Random House, 1987); Robert O. Keohane, *After Hegemony: Cooperation and Discord in the World Political Economy* (Princeton, N.J.: Princeton University Press, 1984).

46 Robert Jervis, *American Foreign Policy in a New Era*, loc. cit., p. 93. On democratic peace, see John Mueller, *Retreat from Doomsday: The Obsolescence of Major* War (New York: Basic Books, 1989).

47 Robert Jervis, "The Era of Leading Power Peace," in Robert J. Art and Robert Jervis, *International Politics: Enduring Concepts and Contemporary Issues*, 7th ed. (New York: Pearson Longman, 2005), p. 407.

48 Jervis, *American Foreign Policy in a New Era*, loc. cit., p. 87.

well have reinforced the administration's conviction that "if America leads, others will follow."[49]

The announced intention of the United States to prevent the rise of any global challenger, and its specific concern to manage if not contain China's rise suggest that hegemonial competition played a role in shaping U.S. policy on North Korea. While the United States took some risk that prodding China to take the lead on the North Korea issue would bolster China's regional influence, officials appear to have calculated that cooperation on the North Korea issue would underscore common security objectives with the United States on the issue and would therefore have a moderating influence on Chinese behavior. They assumed that the DPRK's nuclear designs would ultimately endanger important Chinese security objectives and motivate Beijing to move toward the more hard-line U.S. position—as indeed it has. Should the talks fail, the United States stood to gain from any Chinese-DPRK estrangement and resulting loss of Chinese prestige. Souring relations between Pyongyang and "Big Brother" in Beijing buttressed the assumptions of hard-liners.

While hegemony offers opportunities, it also brings constraints. As the global hegemon, the United States now finds itself tied down militarily and diplomatically by the global war on terror. Cooperation among some rogue states and Islamic extremists conjures up the kind of balancing that has so far failed to emerge among the great or middle powers. The resulting asymmetrical conflict has forced the United States to re-calibrate priorities and to re-allocate resources in ways that have constrained its choices in North Korea. In the run-up to war in Iraq, President Bush told aides, "We do not need another crisis [with North Korea] now."[50] The same logic endured long past the war. Not only did Iraq remain an overwhelming preoccupation, but the DPRK's nuclear breakout created a new reality on the ground in Northeast Asia.[51] U.S. policy continued to shift away from preventing the DPRK from becoming a full fledged nuclear power to preventing transfer of nuclear materials and know-how to other states and groups. PSI was one reflection of this shift. Even before the nuclear tests, administration policy on North Korea was constrained by recognition that a military solution to the North's nuclear program would have catastrophic consequences.[52]

---

49 Ivo H. Daalder and James V. Lindsay, *America Unbound: The Bush Revolution in Foreign Policy*, (Washington, D.C.: Brookings, 2004), p. 44. Daalder and Lindsay note that European reluctance to make the United States pay any penalty for the behaviors they complained of taught the United States that it did not need to heed their complaints. *Ibid.* p. 73.

50 Doug Struck and Glenn Kessler, "Hints on N. Korea Surfaced in 2000," *Washington Post*, October 19, 2002, p. 4.

51 Yonhap News Agency reported on November 28, 2006, that Kim Kye-gwan told reporters "We are ready to boldly participate in the (six way) talks at any time, as we have taken all the defensive measures through the nuclear test." Cited in *NAPSNET Daily Report*, November 28, 2006.

52 "Vice President's Remarks at a Rally for Expeditionary Strike Group 8," Norfolk, Virginia, July 7, 2006 online at <http://www.whitehouse.gov/news/releases/2006/07/20060707-7.html>.

The United States has been constrained in pursuing its objectives by its reliance upon others, particularly China and South Korea, whose interests diverge in significant ways from its own. With its resources tied down on other fronts in the war on terror, and with only limited economic leverage over North Korea, the United States required the cooperation of the North's neighbors and economic partners. However, even in the wake of the DPRK's nuclear test, China and South Korea continue to attach a relatively higher priority to preserving stability on the Korean peninsula than the United States.[53] They also attach a higher priority to managing or even constraining U.S. influence on the Korean peninsula, while the United States seeks to maintain its influence there. These differences had long prevented the United States from pursuing the pressure side of its policy as vigorously as it would have preferred.

U.S. participation in the six-party talks pragmatically recognized the constraints under which it was operating in Northeast Asia, but it also reflected the importance the United States attached to burden sharing.[54] The very power that has created a putative unipolar system gives the United States leverage to pressure others to contribute to collective goods like security. Should it become a reality, CVID would entail enormous implementation costs.[55] The February 2007 Beijing agreement distributes the burden of implementation among the United States and its partners. It is an economic and political boon to the United States if today's stakeholders became tomorrow's burden-sharers.

Alliance politics The United States has had to be mindful of different strategic interests and perspectives of its two allies, Japan and South Korea. Relations with the ROK have been especially sensitive. In the cases of the ROK, different strategic priorities have led to severe strains, despite the desire of both to retain a close relationship.[56] South Korea has faced a classic "entrapment/abandonment" dilemma. Furthermore, its estimate of the economic and political costs of sanctions is far greater than it is for the far-distant United States.[57] As a result, the United States has had to reconcile the goals of increasing pressure on the DPRK with reassuring the ROK of its continued dependability as an ally. Differences have compelled the United States to make concessions to the South at the six-party talks, to reduce the

---

53 "Next Steps on North Korea," colloquium sponsored by the Carnegie Endowment for International Peace, Washington, D.C., October 12, 2006, accessed via the Lexis-nexis database.

54 Joseph S. Nye, "The Battle Between Unilateralists and Multilateralists," in Robert J. Art and Robert Jervis, *International Politics: Enduring Concepts and Contemporary Issues*, 7th ed. (New York: Pearson Longman, 2005), p. 477.

55 Brent Choi, "North Korea: 2005 Outlook," Nautilus Institute, *Policy Forum Online 04-56A,* December 23, 2004 online at <http://www.nautilus.org/fora/security/04-58A_Choi.html>. Cirincione and Wolfstahal, "No Good Choices," loc. cit., p. 276.

56 See Seung-Ho Joo, "US-ROK Relations," in Tae-Hwan Kwak and Seung-Ho Joo, eds., *The United States and the Korean Peninsula in the 21st Century* (Aldershot: Ashgate, 2006), pp. 39-60.

57 Ruediger Frank, "The Political Economy of Sanctions against North Korea," loc. cit., p. 19.

U.S. military footprint in Korea, and to confront the reality that the South would be reluctant to support sanctions, let alone military force, against North Korea.[58] It is in this context that the growing political vulnerability of President Roh Moo-hyun and the Uri party in South Korea have taken on increased significance. Widespread defeat of Uri Party candidates in local elections in 2006 gave the Bush administration hope that a new and more hard-line government might take control of the Blue House before the end of Bush's term.[59] Should the opposition Grand National Party (GNP) form the next government, North Korea would lose another valuable trump card, though it is unclear the extent to which a GNP victory would lead to fundamental policy changes. Even without a change of government, however, Roh's political weakness has already yielded benefits for U.S. diplomacy. In the wake of the twin North Korea "shocks," the government felt compelled to impose limited sanctions and announcing a "review" of its engagement policy.[60]

In contrast to U.S.-ROK differences, U.S.-Japan relations have exhibited increasing harmony of interests, allowing the United States to place greater reliance on that country, through its participation in the PSI, its willingness to impose unilateral sanctions, its more robust policy on confronting North Korea in the Sea of Japan, and its decision to participate in deploying a theater missile defense. Even Japan's nationalist rhetoric about the need for a pre-emptive strike capability against the DPRK serves to remind China of why it is in its interest to end the North's WMD and missile threats. If United States should grant the negative security assurances demanded by the North, both Japan and the ROK might perceive such an action as a sign of waning U.S. commitment to the alliance.[61] Secretary Rice's emphatic affirmation in October 2006 of "the full range" of its treaty commitments was designed to address any abandonment fears, as well as to warn the DPRK.

Issue linkages As a self-perceived hegemon, the United States is more likely than other states to confront issue linkages arising from its multiple commitments and

---

58 According to one report, the ROK reportedly threatened the US that it would have to rethink the alliance if the United States did not make concessions at the September 2005 six-party talks. "S. Korea Pushed US to Give in to NK," *Korea Times*, September 28, 2005 online at <http://times.hankooki.com>. See also David E. Sanger, "Yes, Parallel Tracks to North, but Parallel Tracks Don't Meet," *New York Times*, September 20, 2005 online at <www://nytimes.com/2005/09/20/politics/20assess.html>.

59 Lee Joo-hee, "Opposition fears widening fissure with Washington," *Korea Herald*, May 11, 2006 online at <http://www.koreaherald.co.kr>. Reacting to what he perceived as America's "runaway ally" in South Korea, Nicholas Eberstadt recommended that the United States pursue the same policy in South Korea as it was pursuing in Iran and North Korea, namely "speaking over the heads" of South Korea's appeaser government "directly to the Korean people, building and nurturing the coalitions in South Korean domestic politics that will ultimately bring a prodigal ally back to its senses." Nicholas Eberstadt, "A Real and Present Danger," *American Enterprise Online*, 2005 loc. cit.

60 Ryu Jin, "UN Seeks Bold Action on North Korea," *The Korea Times*, October 9, 2006 online at <http://times.hanooki.com>.

61 Michael Horowitz, "Who's Behind That Curtain? Unveiling Potential Leverage over Pyongyang," *Washington Quarterly*, 28 (1) (Winter 2004/05), p. 26.

involvements. From the time of the so-called "first nuclear crisis" in 1993-94, ending the DPRK's nuclear program was linked to the success of the global nonproliferation regime, U.S. alliance relationships, and increasingly, management of China's rise. In the wake of 9/11 most international security issues, including the North Korea nuclear issue, were drawn into an all-encompassing war on terror. The Bush doctrine, the New National Security Policy and nuclear targeting doctrine signaled a shift in U.S. perception of the appropriate means to deal with North Korea from containment and toward preemption and interdiction.

U.S. policy toward the DPRK has been intertwined with policy toward other rogue regimes, specifically Iran, Iraq, and Libya. To a degree, these linkages have been inadvertent or unintended. Deteriorating situations in Iran and Iraq may have influenced the willingness of the United States to offer concessions in the six-party talks.[62] But they may just as well have stiffened US resolve. Iraq has remained "the central front in the war on terror," and administration officials have openly expressed fear over the impact failure in Iraq would have on the chances to rein in rogue states and terrorists elsewhere. A secondary objective of U.S. diplomacy toward North Korea was "to prevent North Korea's proliferation from becoming a 'how-to-guide' for other countries such as Iran."[63] This concern was given great urgency by the DPRK's nuclear test. By the same token, the United States has been wary of how its posture toward Iran might influence its bargaining position with respect to the DPRK. This linkage has complicated U.S. efforts to come to grips with granting the right to a civilian nuclear program or holding one-on-one talks.[64] Simultaneous nuclear challenges from both North Korea and Iran surely strengthened administration arguments for tough action at the same time that they complicated those efforts.

While inadvertent issue linkages have often complicated diplomacy, the Bush administration has deliberately sought to create a "cross issue linkage" in attempting to sell North Korea and Iran on the Libya paradigm for eliminating their WMD programs. It hoped that the Libya example would provide an object lesson to others of what happens when states "make the right choice" between defiance and cooperation on the issue of WMDs.[65] At the same time, the administration believed

---

62 Joseph Kahn and David E. Sanger, "U.S.-Korean Deal on Arms Leaves Key Points Open," *New York Times*, September 20, 2005 online at <http://www.nytimes.com/2005/09/20/internatinal/asia/20korea.html>. Many believe the administration is correct, but that U.S. troubles in Iraq have already "emboldened these others," Kenneth Katzman, quoted in Farah Stockman, "As Crises Erupt, Sea Changes Seen in Bush Foreign Policy: Less Toughness, More Restraint," *Boston Globe*, July 14, 2006, p. A11.

63 Cirincione and Wolfstahl, "No Good Choices" loc. cit., p. 270. "North Korea's Plutonium Pile Attracts Iran," *The Sunday Times*, January 29, 2006 online at <http://www.timesonline.co.uk/article/0,,2089-2014464,00.html>; Wade L. Huntley, "Goliath's Game: U.S. Policy toward North Korea in Strategic Context," Presented at the workshop "America in Question: Korean Democracy and the Challenge of Non-Proliferation on the Peninsula," Seoul, ROK, May 10-11, 2005, p. 7.

64 Zhiqun Zhu, "North Korean Missiles: Crisis or Opportunity?" *AmericanDiplomacy.org* online at <http://www.unc.edu/depts/diplomat/item/2006/0709/zhu/zhu_nkorea.html>.

65 In a speech at the National Defense University, February 11, 2004, Bush said, "continuing to seek those weapons will not bring security or international prestige, but only

that the war against Iraq would provide instruction in what happens when states "make the wrong choice." The example of Libya's disarmament was particularly important because it appeared to confirm the disarm-first-rewards-later model supported by the administration.[66]

The United States has also been influenced by its partners' use of linkages. Had the United States not accepted the 2005 joint statement, South Korea threatened serious consequences for the alliance, and China made it clear that blame for failure would fall squarely on the United States. Had that happened, any chances that the United States would succeed in enlisting China, Russia, and South Korea in sanctions regime against North Korea would have been severely set back. The United States, not the DPRK, would have been isolated.

Sanctions theory Under the assumptions of neorealism, "the international arena tends to be associated with anarchy and negative sanctions."[67] Taking as a baseline Elliott and Hufbauer's "nine commandments" for applying negative sanctions, the North would have seemed a promising target.[68] The economy is small and, despite *Juche*, highly dependent on outside trade, tourism, aid, and transfers, as well as upon illicit activities such as counterfeiting and drug trafficking. While it is acutely dependent upon a few much larger economies, those economies rely little upon resources from or trade with it. On the other hand, the cost of sanctions to the United States is quite low. The Bush administration's enthusiasm for sanctions, and confidence that they would bring about the demise of the regime, can be explained in terms of this logic. The theoretical arguments for smart sanctions, in particular financial sanctions, provides a rationale for the administration's preference for them.[69] The fact that North Korea has been able to exploit moral hazard to insulate itself against sanctions would have reinforced the administration's conviction that the chief obstacle to effective sanctions lay with China and South Korea. Their reluctance has meant that the United States has been unable to impose sanctions "quickly and decisively" or comprehensively as required in theory.[70]

---

political isolation, economic hardship and other unwelcome consequences." See Robert S. Litwak, "Non-proliferation and the Dilemmas of Regime Change," loc. cit., p. 18.

66 Peter Hayes et al., "Light Water Reactors at the Six-party Talks: The Barrier that Makes the Water Flow," loc. cit.

67 David A. Baldwin, *Paradoxes of Power* (New York: Basil Blackwell, 1989), p. 70.

68 Gary Clyde Hufbauer, Jeffrey J. Schott, and Kimberly Ann Elliot, *Economic Sanctions Reconsidered: History and Current Policy* (Washington, DC: Institute for International Economics, 1990), pp. 94-105. On the other hand, the North Korea situation fails to conform to their findings that the goal should be relatively modest and that the target and sender enjoy friendly relations.

69 See Ruediger Frank, "The Political Economy of Sanctions against North Korea," *loc. cit.*, p. 16.

70 Ruediger Frank, "The Political Economy of Sanctions against North Korea," loc cit., p. 13. Michael Horowitz contends that there are means of choking off the North's access to hard currency that could be successful even without South Korea's participation. Horowitz, "Who's Behind That Curtain?" loc. cit., p. 39.

Despite evidence (both theoretical and experiential) that threats generate resistance to an influence attempt, the administration long resisted positive sanctions.[71] The preference for negative sanctions implied acceptance of the utility of making North Korea feel threatened, and hence, more inclined to make concessions. Further, this preference implied rejection of prospect theory's argument that driving the target into the realm of losses would make it more likely to act aggressively than it would otherwise have done.[72] The Bush administration's commitment to deferring rewards is consistent with Schelling's "compellant threat," in which punishment is continued until the target has completed the desired action, rather than making fulfillment of the threat contingent upon failing to comply.[73]

State Level Factors

State level theory assumes the importance of regime type, internal actors, political and social institutions and processes, culture, and capabilities in shaping the international behavior of states.

Political Culture

The Bush foreign policy lies in many respects well within the parameters of American political culture. Its Manicheanism, exceptionalism, evangelical universalism, and elements of both realism and liberal internationalism can all be found throughout the history of the republic. The current administration's embrace of democracies and open contempt for "tyrants" and "rogues" echoes the cold war divide between the free and the not-free as well as the founding doctrine of the "two spheres" distinguishing America from what was seen as a cynical and decadent Europe. The United States' pessimism about the DPRK's international behavior is shaped not only by its past record, but by the widely shared belief that this is the way dictatorships behave.[74] Despite the strain of crusading militancy, however, U.S. diplomacy toward North Korea has continued to reflect the realist strain of United States foreign policy tempering power with prudence. To date, U.S. policy behavior toward North Korea has been characterized more by caution than by recklessness.

Factional Politics and Two Level Games

It has become something of a truism that the Bush foreign policy has been plagued by a wide ranging battle over foreign policy between a coalition of hard-liners and another of pragmatists. This is no less true of policy on North Korea. Policy inconsistencies have reflected that struggle. In this conflict, the argument has

71 Michael Horowitz, "Who's Behind That Curtain?" loc. cit., p. 74.

72 SeeVictor D. Cha, "Hawk Engagement and Preventive Defense on the Korean Peninsula," *International Security*, 27 (1) (Summer 2002), pp. 40-78.

73 Cited in Baldwin, *Paradoxes of Power*, loc. cit., p. 49.

74 Jervis, "Explaining the Bush Doctrine," in Robert Jervis, *American Foreign Policy in a New Era*, loc. cit., pp. 80-81.

centered on a contest pitting the goals of "regime change against the goal of behavior change."[75] In preparing to deal with the North Korea problem, the "pragmatic realists," located principally in the State Department and National Security Council, expected to pick up where the Clinton administration had left off, with bilateral negotiations for improved implementation of the Agreed Framework.[76] After the collapse of the Agreed Framework, they supported renewed negotiations, albeit in the six-party format, and a degree of flexibility in the U.S. position. Their influence was visible most clearly in the September 2005 Joint Statement and again in the February 2007 Beijing agreement.

Another group consisting of officials in the Vice President's office, the Defense Department, and non-proliferation specialists such as John Bolton and Robert Joseph in the State Department, advocated a difference course.[77] An alliance of neoconservatives and "assertive nationalists" linked a favorable outcome in North Korea to the muscular assertion of U.S. hegemony in removing dangerous enemies while bringing democracy to benighted regions.[78] They believed that "our post-Vietnam reluctance to use military force ... sent a message of appeasement and peace at any price to opponents around the world," while Bush administration "victories in Iraq and Afghanistan demonstrate that the United States cannot be assumed to be a paper tiger."[79] This group held the upper hand at least through the end of 2005. Scorning the Agreed Framework, they advocated a policy of regime change through pressure and isolation as the best means to achieve U.S. security objectives.[80] Some senior administration officials apparently entertained as late as 2003 the viability of a military solution.[81] The influence of the second group was evident

75 Litwak, "Non-proliferation and the Dilemmas of Regime Change," *loc. cit.*), p. 18, 26.

76 Daalder and Lindsay, *America Unbound*, loc.cit, pp. 45-46. Krauthammer uses the term "pragmatic realists." Charles Krauthammer, "The Unipolar Moment Revisited," *The National Interest*, 70 (Winter 2002/2003).

77 The appointment to the NSC of Victor Cha, an academic best known for his deep skepticism of the Kim regime, was consistent with the increase of the pressure track after the 2004 election, but his writings on hawk engagement suggested greater openness to serious negotiations, if only to provide a genuine test of their intentions.

78 For a comparison of the views of neoconservatives and assertive nationalists, see Daalder and Lindsay, *America Unbound,* loc. cit., pp. 15-16. For a highly critical view, see also, Leon V. Sigal, "The Cabal Is Alive and Well," Nautilus Institute, *Policy Forum Online 05-95,* November 29, 2005 online at <http://www.nautilus.org/fora/security/0595Sigal.htm>.

79 Gordon Cucullu, "A Stiff Test for America," American Enterprise Online, 2005 online at <http://www.taemag.com/issues/articleID.18596/article_detail.asp>.

80 Larry Niksche, *North Korea's Nuclear Weapons Program,* Congressional Research Service, CRS Issue Brief for Congress, IB91141, August 31, 2005, p. 5; Robert S. Litwak, "Non-proliferation and the Dilemmas of Regime Change," loc. cit., p. 8; Ruediger Frank, "The Political Economy of Sanctions Against North Korea," loc. cit., p. 29.

81 Ted Galen Carpenter and Doug Bandow, *The Korean Conundrum: America's Troubled Relations with North and South Korea* (New York: Palgrave McMillan, 2004), p. 82. North Korea would certainly have been aware of the charges by Seymour Hersh that the administration was actively planning military action against another other member of the axis of evil, Iran. Seymour Hersh, "The Iran Plans," *The New Yorker*, April 17, 2006

almost immediately in public repudiation of South Korea's "sunshine policy" and rejection of continuing the Clinton-era dialogue as "deeply, profoundly misguided and wrong."[82] Pressure from hard-liners severely constrained the freedom of the Bush administration's first negotiator, James Kelly.[83] Hard liners largely succeeded in setting the parameters within which the six-party talks took place, for example in drawing the line against accepting a freeze as a step toward dismantlement and in interpreting the Joint Statement.[84] Even on the eve of the February 2007 session of the six party talks, Senator Richard Lugar still expressed concern about officials who were creating "countervailing situations" that were undermining US diplomacy.[85]

Administration defenders have called the two-track policy of pressure and negotiation "effective harmony of different goals," and a successful reconciling of the two warring camps.[86] However, others have taken a different view, some charging that the result has been incoherence, a policy that is neither fish nor fowl, neither engagement nor confrontation.[87] Hathaway writes that,

> In no area has the ideological struggle had more unfortunate consequences than with respect to the contradictions between the administration's repeated insistence that it will not tolerate a nuclear North Korea, and the startling absence of action to head off this possibility ... there has been a real disconnect between the very real dangers posed by an active North Korean weapons program and the absence of a concerted policy to counter these dangers.[88]

Nicholas Eberstadt, one of the most consistent hawks on the North Korea issue, noted that hard-liners had been scuttling U.S. diplomatic efforts: "Convinced that diplomacy will be futile—or perhaps worried that it might actually succeed, and

---

online at <http://www.newyorker.com/fact/content/articles/060417fac_fact>; and, "Watching Lebanon: Washington's Interest in Israel's War," *The New Yorker*, August 28, 2006, pp. 28-33.

82 Kurt Campbell in "North Korea's Nuclear Test," a colloquium sponsored by the Center for Strategic and International Studies, October 11, 2006 retrieved from the Lexis-nexis database, November 2, 2006.

83 Murray Hiebert and Gordon Fairclough, "North Korea Talks' Duration Signals New Tacks," *Wall Street Journal*, August 1, 2005, p. A7.

84 Daryl G. Kimball "Getting Serious about North Korea," *Arms Control Today*, December 2005 online at at <http://armscontrol.org/act/2005_12/focus.asp>; C. Kenneth Quinones remarks, "U.S.-North Korean Relations," loc. cit.

85 U.S. Congress, Senate, Committee on Foreign Relations, "Nomination of John Negroponte to be Deputy Secretary of State," Hearings, January 30, 2007, Federal News Service, Lexis-Nexis database.

86 Michael Green, cited in Paul Kerr, "North Korea, U.S. Talks Inch Forward," *Arms Control Today*, April 2005 online at <http://www.armscontrol.org/act/2006_04/nkustaklsinch.asp>.

87 Robert J. Einhorn, "The North Korea Nuclear Issue: The Road Ahead," Nautilus Institute, *Policy Forum Online* 04-33A, September 14, 2004 online at <http://www.nautilus.org/fora/security/0433A_Einhorn.html>.

88 Robert Hathaway, "Supping with the Devil" [book review], *World Policy Journal*, Winter 2003/2004, p. 85.

thereby give the North Korean regime a new lease on life—these officials have little incentive to fight for politically unpopular compromises that might lead to a workable agreement."[89] Isolation, not an Agreed-Framework type of agreement was the objective, and whenever the talks appeared to be alleviating the pressure, hard liners were always able to rein them in. In 2004 at an early stage in the six-party talks, Vice President Cheney is reported to have circumvented the State Department and convinced President Bush to quash an emerging document "that lacked the tough language on disarmament ... that Mr. Cheney knew that Mr. Bush wanted."[90] It appeared that by the end of 2005, the hawks had come into the ascendancy.[91] The restrictive exegesis of the 2005 Joint Statement was reportedly drafted by hawks without input from Christopher Hill.[92] Treasury Department sanctions on eight DPRK companies in October may have contributed to the subsequent suspension of talks, but hawks rejoiced that the United States had finally found a way to inflict real pain on the North's elites.[93] The new U.S. ambassador to the ROK ensured that Washington hawks' views would remain at the table when he revived charges that North Korea was a "criminal regime" whose counterfeiting activities were reminiscent of the Hitler regime.[94] North Korea's missile and nuclear tests appeared to prove hard liner's longstanding conviction that the path of negotiation was fruitless, that the Joint Statement was an appeasement document, and that the North would only respond to firmness and pressure.

Despite their skepticism regarding any credible agreement with the DPRK, hawks have not been entirely opposed to negotiation. An original rationale for hawk engagement was to enlist Washington hard-liners' support by using the argument that "exhausting all diplomatic options" would better secure support for a tough

89 Ibid., p. 88.

90 David E. Sanger and Eric Schmidt, "Cheney's Power No Longer Goes Unquestioned," *New York Times*, September 10, 2006, p. A1, 22; Leon V. Sigal, "What North Korea's Missile test Means," Nautilus Institute, *Policy Forum Online* 06-62 July 27, 2006.

91 Guy Dinsmore and Anna Fifield, "US hardliners grab North Korea policy reins," *Financial Times*, December 19, 2005 online at <http://www.ft.com/cms/a/92d94ba6-24e4-11d8-81c6-08209b00dd01,gaid=05219006073>; Leon V. Sigal, "The Cabal is Alive and Well," Nautilus Institute, *Policy Forum Online* 05-95A, November 29, 2005 online at <http://www.nautilus.org/fora/security/0595Sigal.htm>; Gavan McCormack, "Criminal States: Soprano vs. Baritone—North Korea and the US," *Japan Focus* (2006) online at <http://www.japanfocus.org/products/topdf/2189>.

92 Charles L. Pritchard, "Six-party Talks Update: False Start or a Case for Optimism," presented at The Changing Korean Peninsula and the Future of East Asia, Center for Northeast Asian Policy Studies, the Brookings Institution Seoul Forum of International Affairs, December 1, 2005, p. 7

93 Joel Brinkley, "U.S. Squeezes North Korea's Money Flow," *New York Times*, March 10, 2006 online at <http://www.nytimes.com.2006/03/10/politics/10korea.html>. James Cotton, "Missiles and Sanctions: Has a Watershed Been Reached in the North Korean Nuclear Crisis?" Nautilus Institute, *Policy Forum Online* 06-75, September 13, 2006.

94 James Brooke, "Talks Stalled, U.S. Envoy Matches Insults of North Korea," *New York Times*, December 15, 2005 online at <http://www.nytimes.com/2005/12/15/international/asia/15seoul.html>.

policy later.[95] Only if the United States could take the high ground and demonstrate convincingly that it had given North Korea a fair chance to "make the right choice," would it be able to pursue a policy of sanctions or force that had both domestic and international support. The United States' apparent willingness to subscribe to the 2005 Joint Statement and subsequently to endorse it in SC Res 1718 was consistent with such a motivation.[96]

Notwithstanding the apparent ascendancy of hawks, recent U.S. actions on the diplomatic front suggest a substantial turnaround in the fierce "ideological war" within the administration.[97] The departure of influential hawks like Donald Rumsfeld, John Bolton and Robert Joseph following Democratic congressional victories opened the possibility that the domestic political balance might shift in favor of the pragmatists, assuming that President Bush was prepared to move in their direction.[98] This possibility seemed confirmed by the dramatic U.S. diplomatic offensive at the six party talks at the end of 2006 and early in 2007. Having long resisted front-loaded incentives as "rewarding bad behavior," the United States incorporated costly concessions into the Beijing agreement. These even included willingness to repeal or modify the very financial sanctions that hawks believed the US's most effective tool. The allergic reaction to the agreement expressed from the right appeared to confirm that it represented a genuine change of tack by the administration.[99]

Congress

For six years, the Republican congress conducted a guerrilla war against the Agreed Framework, which it regarded as "deeply flawed" and a case of "appeasement and bribery."[100] After G.W. Bush's election in 2000, Congress continued to be an influential player on the sidelines, ensuring through hearings, resolutions and legislation that the spotlight would continue to illuminate the evils of the Kim Jong-

95 Victor D. Cha, "Korea's Place in the Axis," *Foreign Affairs*, 81 (3) (May/June 2002), p. 81, 89.

96 B.C. Koh, "Six-party Talks: Round 3," Nautilus Institute *Policy Forum Online 04-26*, July 1, 2004 online at <www.nautilus.org/fora/security/0426A_Koh.html>.

97 David E. Sanger, "Outside Pressures Broke Korean Deadlock," *New York Times*, February 14, 2007, p. A-10.

98 See, for example, "David E. Sanger and Scott Shane, "In Gates Selection, White House Hopes to Close Rift Between State and Defense," *New York Times*, November 12, 2006 online at <http://www.nytimes.com.2006/11/12/washington/12gates.html>. Sanger and Shane note, however, that during the first North Korea crisis in 1994, Gates gave qualified support to missile strikes on DPRK nuclear facilities.

99 Bill Gertz, "Bolton Hits Agreement as 'Bad Signal' to Iran," *Washington Times*, February 14, 2007. "Online News Hour, February 13, 2007 <http://www.pbs.org/newshour/bb/asia/jan-june07/koreadeal_02-13.html>; Helene Cooper and Jim Yardley, "Pact with North Korea Draws Fire from a Wide Range of Critics," *New York Times*, February 14, 2007, p. A-10.

100 The language was used in a Republican task force report at the end of 1999. Ted Galen Carpenter and Doug Bandow, *The Korean Conundrum: America's Troubled Relations with North and South Korea* (New York: Palgrave MacMillan, 2004), p. 55.

Il regime. Congressional gains in 2002 strengthened the administration's hand as it confronted the new crisis in Korea. The 2004 North Korean Human Rights Act, for example, authorized $24 million per year toward to promote human rights and democracy in the DPRK, and Rep. Porter Goss recommended using undercover agents to penetrate the government. Republican gains in the 2004 elections further strengthened the influence of the skeptics of engagement. The End Dictatorship, Assist Democracy Act of 2005 set a goal of transforming 45 autocracies, including the DPRK through nonmilitary means. Congress joined administration hawks in condemning the September 2005 Joint Statement, and in October 2006, the North Korea Non-Proliferation Act urged "all governments to comply with S.C. Res. 1695." Congress' greatest leverage has been not so much in what it has done, but in what it would not consent to do. Since 1995, Congress has made it clear that it would refuse to approve or fund any proposal for significant U.S. aid to or security assurances for the DPRK. Though generally espousing the agenda of administration hawks, the Republican congress was reluctant to get too far ahead of the Bush administration on the North Korea issue.[101]

Congress' hard line served the administration well in negotiations by "tying the administration's hands" when it came to pressure to make concessions. Even where Bush has failed to press the Right's agenda vigorously enough, Republican control of Congress insulated the administration against congressional criticisms that surely would have dogged a Democratic president had he pursued some of the very actions that Bush has pursued.[102] Intriguingly, even before the 2006 elections, congress was becoming more assertive on a number of foreign policy questions, including North Korea, and not necessarily in a hawkish direction.[103] By mandating appointment of a North Korea policy coordinator, an office first established by the Clinton administration and later dropped by the Bush team, congress further demonstrated a more independent streak. The stunning Democratic election victory in 2006 could be expected significantly to bolster support for a policy closer to Clinton-style engagement formerly eschewed by congress. On the other hand, Congress' position on human rights issues was less likely to change, and it is far from certain that the Democratic congress would automatically approve the president's funding requests pursuant to the Beijing Agreement. In general, however, there was a strong

---

101 Karin J. Lee, "The North Korean Human Rights Act and other Congressional Agendas," Nautilus Institute, *Policy Forum Online* 04-39A, October 7, 2004 online at <http://www.nautilus.org/fora/security/ 0439A_Lee.html>.

102 Robert Hathaway, "Supping with the Devil" [book review], *World Policy Journal* (Winter 2003/2004), p. 89.

Paul J. Saunders and Nicholas K. Gvosdev, "Pyongyang and America Priorities," *In the National Interest* 2, 2 (January 2003) online at <http://www.inthenationalinterest.com/articles/Vol2Issue2/Vol2Issue2Realist.html>.

"To Pyongyang Via Baghdad," *Wall Street Journal*, January 13, 2003 via Proquest database June 1, 2006.

103 John Feffer, "North Korean Fireworks?" Nautilus Institute, *Policy Forum Online* 06-52, July 3, 2006 online at <http://www.nautilus.org/fora/security/0652Feffer.html>.

possibility that Congress would have little choice but to remain largely reactive to the choices made by the nation's chief diplomat.[104]

Media and Think Tanks

Conservative commentary has been harshly critical of the DPRK and generally supportive of sanctions or even the use of force. In January 2003, prior to the Iraq war, Paul J. Saunders wrote in *The National Interest* that the United States should consider a military strike on North Korea before becoming engaged in Iraq.[105] The *Wall Street Journal* also favored military pressure, but not until after disposing of the Iraq threat. "Toppling Saddam with dispatch," argued the *Journal*, "would allow the U.S. to turn its military attention away from the Gulf and toward the crisis in Korea."[106] After three years of the status quo, however, conservative critics began to lose patience with an administration whose bark, they thought, continued to be far worse than its bite. The *Weekly Standard* accused Bush of "Clintonian" temporizing.[107] Nicholas Eberstadt at the American Enterprise Institute excoriated the administration for being "unwilling to move against North Korea's nuclear challenges by itself," and accused it of a "climb-down" that was "dizzying to watch."[108] Criticism from the right reached a crescendo following the February 2007 Beijing agreement, when the *Wall Street Journal* scorned the agreement as "faith-based nonproliferation."[109] Even Ashton Carter, Clinton's Assistant Secretary of Defense and William J. Perry, his Defense Secretary, concluded prior to the DPRK missile tests that "diplomacy has failed," and recommended that the United States should "strike and destroy" its missiles on the launch pad.[110] Though Bush has prided himself on not paying attention to media critics, such criticism was symptomatic of eroding political support for the administration's policy among foreign policy elites.

---

104 For an interesting discussion of alternative policy consequences of the Democratic Party victory in the elections, see Moo-jin Yang, "The Democratic Party's Victory in the U.S. Midterm Elections and the North Korean Nuclear Issue," Nautilus Institute, Policy Forum Online 06-101A, December 5, 2006 online at <http://www.nautilus.org/fora/security/06101Yang.html>.

105 Paul J. Saunders and Nicholas K. Gvosdev, "Pyongyang and America Priorities," In the National Interest 2, 2 (January 2003) online at <http://www.inthenationalinterest.com/articles/Vol2Issue2/Vol2Issue2Realist.html>.

106 "To Pyongyang Via Baghdad," *Wall Street Journal*, January 13, 2003 via *Proquest* database May 22, 2006.

107 Michael Rubin, "Bubba Dubya?," *Weekly Standard* 011, 38 (June 19, 2006); William Kristol, "Kim's Choice—and Bush's," *Weekly Standard* 001, 41 (July, 2006) online at <http://www.theweeklystandard.com>.

108 Nicholas Eberstadt, "Nuclear Shakedown," *Wall Street Journal*, July 6, 2006 ProQuest database online at <http:proquest.umi.com>; and "Kim Jong-Il's Nuclear Ambitions," Nautilus Institute, *Policy Forum Online* 07-010A, February 6, 2007 online at <http://www.nautilus.org/fora/security/07010Eberstadt.html>.

109 "Faith-Based Nonproliferation," editorial, *Wall Street Journal*, February 14, 2007.

110 Ashton B. Carter and William J. Perry, "If Necessary, Strike and Destroy," *Washington Post*, June 22, 2006, p. A29.

Public Opinion and Electoral Politics

There is little evidence that the Bush policy toward North Korea has been shaped by public opinion. North Korea was a peripheral issue in the 2000 campaign, and the 2002 congressional elections were over by the time of the second nuclear crisis. John Kerry made North Korea an issue in 2004, charging that administration engagement was too little, too late, but attacks by the liberal Kerry were hardly likely to force the Bush campaign to shift positions. The United States had already moderated its position enough at the June 2004 session of the six-party talks to argue that it was negotiating seriously, but Bush had to be more concerned about his anti-appeasement base.

By defining all aspects of U.S. security policy in terms of a war on terror, the administration ensured that security in Northeast Asia and Korea would be perceived through the lens of homeland security.[111] A large majority of U.S. public opinion consistently has viewed the DPRK as a threat to the United States, and could therefore have been expected to be permissive regarding a tough administration policy. However, the public continued, even after the DPRK's nuclear test, to oppose using military means to end the threat and to favor a diplomatic solution.[112] Growing public dissatisfaction with the Bush Iraq policy also implied reduced latitude for pursuing the military option in other cases. Barring new terrorist horrors or even worse surprises from the DPRK that could reignite public enthusiasm for military action, domestic politics reinforces other factors militating against any military strike on the DPRK.

Individual Level

In international relations theory, the individual level of analysis generally refers to the role of individual human actors in shaping state behavior.[113] It stresses psychological variables such as personality, perception and cognition, as well as individual and socially mediated attitudes and values. Though it is often difficult to disentangle individual level influences from state and even system level influences, the importance of policymakers' "ideational dispositions" in shaping the Bush foreign policy has been widely remarked.[114] James Mann has written that many of

---

111 David C. Kang, "A Better Strategy," *Washington Post*, November 28, 2005 online at <http://www.washingtonpost.com/wp-dyn/content/article/2005/11/27/AR2005112700944_>

112 Gallup Poll, February 4-6, 2002; CBS News Poll, January 4-6, 2003; Newsweek Poll, October 19-20, 2006, at PollingReport.com online at <http://www.pollingreport.com/korea.html>. Fox News/Opinion Dynamics Poll, July 11-12, 2006. *PollingReport.com* online at <http://www.pollingreport.com/korea/htm>.

113 Urie Bronfenbrenner, "Mirror Image in Soviet-American Relations," Ole R. Holsti, "Belief system and natinal images: a case study," *Journal of Conflict Resolution*, 11 (1962), pp. 244-245.

114 Wade L. Huntley, "Goliath's Game: U.S. Policy toward North Korea in Strategic Context," Presented at the workshop "America in Question: Korean Democracy and the Challenge of Non-Proliferation on the Peninsula," Seoul, ROK, May 10-11, 2005.

the key shapers of the Bush foreign policy believed in the centrality of traditional national security issues, the key role of military power, the inherently positive place of American power and ideals in the world, and great optimism about the country's future.[115] From this perspective, the United States must not hold back from using its unprecedented power, alone if necessary, to protect its interests.[116] Furthermore, multilateral institutions and treaties were perceived as unjustly constraining the United States in these noble pursuits.[117] In the neoconservative variant advocated by a small number of influential officials in Bush's first term, the United States had an historic opportunity to transform the world system in ways that would leave the U.S. free of state challengers far into the future and permit at the same time the promotion of universal democratic and liberal values.[118] Such suspicions have been reinforced by the strongly Manichean views expressed in pronouncements of the president and other officials. Perceiving the United States as virtuous and its adversaries as evil, the administration has resisted policies of engagement that are believed to prolong the life of dictatorial regimes: Better to isolate and strangle them.

Strong personal perceptions have shaped the policy debate on North Korea from the start of the Bush administration. The pronounced negative enemy image and open contempt for Kim Jong-Il, the near allergic reaction to Clinton's engagement, and the reluctance to talk directly with what was regarded as an evil regime often gave the appearance of reflecting personal dispositions rather than dispassionate calculations of state's interest. A number of central actors believed that North Korea would be compelled to give in to U.S. demands or face imminent collapse.[119] Since the authoritarian nature of the regime was believed to be the source of its aggressive behavior, it was logical to conclude that negotiation with "inveterate cheaters" was fruitless and that only regime change could end that behavior.[120] Bush administration officials' characterizations of the DPRK mirror the "inherent bad faith" enemy image once associated the Soviet Union during the Cold War.[121] Policy decisions may have been influenced further by the association of negative sanctions with positive personal traits (e.g. toughness, determination, courage) and of positive sanctions to negative personal traits (e.g. weakness, vacillation, appeasement).[122]

President Bush, Vice President Cheney, Secretary of State Condoleezza Rice, U.S. ambassador to the ROK Alexander Vershbow, U.N. Ambassador John Bolton and others have all publicly heaped scorn on the DPRK and Kim Jong-

115 Mann, loc. cit., p. xvi.

116 Daalder and Lindsay, *America Unbound, loc.cit.*, pp. 40-41. Mann, *Rise of the Vulcans*, loc. cit., p. xvi.

117 Ibid., p. 44.

118 Daalder and Lindsay, *America Unbound*, loc. cit., p. 47.

119 Mann, *Rise of the Vulcans*, loc. cit., p. 280.

120 Woodward, Joel S. Wit, Daniel B. Poneman and Robert L. Gallucci, *Going Critical: The First North Korean Nuclear Crisis* (Washington, D.C.: Brookings, 2004), p. 404.

121 Urie Bronfenbrenner, "The Mirror Image in Soviet-American Relations: A Social Psychologist's Report," *Journal of Social Issues*, 17 (3) (1961), pp. 46-48; Ole R. Holsti, "The Belief System and National Images: A case study," *Journal of Conflict Resolution* VI (1962), pp. 246-247.

122 David A. Baldwin, *Paradoxes of Power* (New York: Basil Blackwell, 1989), p. 76.

Il.[123] By expressing his visceral loathing for Kim Jong-Il, President Bush early on personalized the conflict with North Korea. His scorn for the North Korean leader was complemented by a longing to spread what were perceived to be universal values of freedom and democracy to the unfortunate inhabitants of the DPRK.[124] Like most American presidents before him, though perhaps more energetically than many, he has framed his presidency in terms of acting out the United States' moral mission.[125] Bush's attitude toward the DPRK can be explained in part by his belief in "a duty to free people," and expressions of deep concern to relieve human suffering such as found in North Korea.[126] He appeared to reject warnings about the economic consequences of a collapsed North Korean regime on the grounds that "either you believe in freedom...or you don't."[127] His strong convictions would have been buttressed by a propensity to "stay the course" and to discount evidence that contradicted the wisdom of that course.[128] It has been reported that Bush personally approved the timing of new U.S. financial sanctions to coincide with the issuance of the September 2005 Joint Statement.[129]

Undersecretary for Arms Control and International Security and later U.N. ambassador, John Bolton was a highly-visible hawk, promoting PSI and other sanctions and aggressively lobbying for UN Security Council sanctions. Policy proposals sketched out by Victor D. Cha shortly before he was appointed the National Security Council's Director for Asian Affairs, closely resembled the subsequent actions taken by the administration to increase pressure on the DPRK.[130]

Notwithstanding the personal beliefs of officials, the absence of a strong response when the DPRK broke out of the NPT, as well as its continued faithfulness to the six-party process and willingness to pursue an "early harvest" concept of rewards for provisional steps by the DPRK, failed to conform to the behavior expected from the individual level factors mentioned above. The administration's caution

123 James Cotton, "Whither the six-party process on North Korea?" *Australian Journal of International Affairs*, 59 (3) (September 2005), p. 276.

124 Robert Jervis, American Foreign Policy in a New Era, p. 81; Bob Woodward, *Plan of Attack* (New York: Simon and Schuster, 2004), pp. 88-89.

125 Bob Woodward, *Bush at War* (New York: Simon and Schuster, 2002), p. 130, 340.

126 In interviews with Woodward Bush pointedly said that the need to relieve suffering was a motive in his attitudes toward Afghanistan, Iraq and North Korea. Woodward, *Bush at War,* loc. cit., p. 88, 131, 340, 339, 341.

127 Woodward, *Bush at War*, loc. cit., p. 340.

128 Woodward, *Bush at War*, loc. cit., p. 342.

129 Michael Hirsh, Melinda Liu and George Wehrfritz, "How N. Korea Changed the Nuclear Club's Rules," *Newsweek*, October 23, 2006 online at <http://www.msnbc.com/id/15265432/site/newsweek/from/ET/......>.

130 Victor D. Cha and David C. Kang, "The Debate over North Korea," *Political Science Quarterly*, 119 (2) (2004), p. 251. Though Cha no longer supported engagement of any kind, he sketched out a coercive strategy aimed at rallying "interested regional powers to isolate the regime until it gave up its proliferation threat." "The United States and its allies ... would intercept any vessels suspected of carrying nuclear—or missile-related materials in an out of the North. Secondary sanctions would also be levied against firms in Japan and other Asian countries involved in illicit North Korean drug trafficking in an effort to restrict the flow of remittances to the DPRK leadership."

showed every sign of the kind of pragmatism that one might have expected from any president in light of the material and diplomatic constraints on the United States. The departure of hard-liners Rumsfeld, Bolton and Joseph from the administration in 2006 and 2007 suggested a further shift of individual-level factors toward greater pragmatism.[131] The promotion of the U.S. chief negotiator Ambassador Christopher Hill to the position of North Korea policy coordinator also seemed to bode well for the negotiating track. It has been suggested that Rice herself, an often vocal critic of the DPRK regime, may have acted decisively to push through the 2007 Beijing accord despite vociferous objections both from within and without the administration.[132] Indeed, the president and other members of the administration had stressed repeatedly throughout the autumn of 2006 that the US goal in the six party talks would be to reaffirm and enforce the September 2005 Joint Statement—a statement opposed by hard-liners. With the conclusion of the Beijing Agreement, for the first time since the six-party talks began, tangible progress in reaching that goal seemed at least possible.

Conclusions

Since its inception, the Bush administration's policy has been grounded in a strong preference for isolation over engagement, for pressure over incentives. This preference reflected a shared *Weltanshauung* of individual actors. That policy was constrained, however, by system and state level realities that drove the United States to eschew forceful unilateral action and to pursue its objectives through multilateral regional diplomacy. Initially, the Bush administration focused on demonstrating a decisive break from the Clinton policy, which the administration saw is too eager, too bilateral, and too reliant upon good faith agreements. For a time, the administration let the DPRK "cool its heels" pending a policy review. Following the breakdown of the Agreed Framework, U.S. policy was guided by the principles of CVID, no rewards for bad behavior and no bilateral negotiations. That policy was pursued along twin tracks of multilateral engagement, as evidenced in the six-party talks, and compellance, as evidenced in the Proliferation Security Initiative and pursuit of theater missile defense with Japan. The six-party talks were compatible with the aims of both hawks and pragmatists. For pragmatists, they were *the* game and the only realistic hope to achieve a denuclearized DPRK. For hawks, the six party talks they were but a tactical side-show in a wider strategy of isolation. For them, the talks were both a holding action while the administration sought effective means of exerting pressure on North Korea and a means of mobilizing the DPRK's chief benefactors to seal off the regime. After two years of drift and frustration at the talks, the September 2005 Joint Statement appeared to offer promise of a peaceful resolution, but subsequent actions by both the United States and the DPRK put that prospect in doubt and seemed to confirm that isolation and pressure remained

131 Farah Stockman, "As Crises Erupt, Sea Change Seen in Bush Foreign Policy; Less Toughness, More Restraint," *Boston Globe*, July 14, 2006, p. A 11.

132 David E. Sanger and Thom Shanker, "Rice is Said to Have Speeded North Korea Deal," February 16, 2007, p. A3.

the default position of the United States. The DPRK's defiant missile and nuclear tests confirmed for administration hawks the validity of their preferred policy of pressure and their scorn for pursuing written agreements with rogue states. Despite its modest beginning and uncertain results, PSI provided a template for creating a broader coalition of the willing for exerting pressure on the DPRK, a template the United States was happy to apply with increased vigor following Security Council resolutions 1695 and 1718. In terms of the correlation of forces, the Bush administration's relative ability to inflict harm on the DPRK had never seemed greater.

At the end of 2006, however, the correlation of forces had begun to shift again. A new constellation of individual, state and international level influences was coming into alignment. At the individual level several of the most influential administration hawks had departed, and some of their most aggressive policies had been undermined by events in Iraq and the perceived need to focus US energy on multiple threats from Iran. As a consequence of the changing cast of characters in the executive branch and the shift of congressional power to the Democratic Party, it appeared less likely that the state level would be able to sustain a policy of strangulation of the DPRK. Bush's lame duck status could be expected to complicate his pursuit of any of the options open to him. Just as the objective security threat from the DPRK was increasing, the ability of the Bush administration to "stay the course" was eroding and its bargaining leverage waning. At the system level, the DPRK's concrete demonstration of its deterrent capability, combined with the crises in Iraq, Iran, and Lebanon and on the Israeli/Palestinian front were likely to weaken the hand of the United States. The combination of all these developments created a more level playing field at the six-party talks. There was a danger that both the United States' negotiating partners and its adversaries would take advantage of perceived U.S. weakness. But there was an optimistic scenario as well, namely that the new equilibrium might lead to more flexibility in the U.S. and DPRK positions, as China had been urging. The modest success of the February Beijing Agreement lent a degree of plausibility to that scenario.

It has been argued here that the Bush administration's policy toward the DPRK was more coherent than many critics have charged. It put building a regional consensus for pressuring North Korea ahead of continuing the negotiating path chosen by the Clinton administration. It pursued a strategy of delay which eventually bought more international legitimacy for its hard line, and which eventually led to some progress on the negotiating front, but only at the expense of allowing the DPRK to become a proven member of the nuclear club. In the end, the personal preferences of leaders could not overcome recalcitrant system and state level realities. By the time the US diplomatic strategy for isolating the DPRK had begun to work, the situation in Iraq had deteriorated further, Democrats controlled congress, and, most important, the North Korean nuclear horse had trotted out the barn door. It was far from certain that it could be coaxed back inside.

Chapter 5

China and the North Korean Crisis: The Diplomacy of Great Power Transition

Jeremy Paltiel

The Korean Peninsula and China as a Great Power

The history of China's involvement in the Korean Peninsula marks its fall and rise of China as a great power. China has traditionally regarded Korea as its "lips and teeth." When Japanese under the Toyotomo Hideyoshi invaded Korea in the 1590s the Ming Dynasty nearly bankrupted itself to help the Korean Dynasty to repel the Japanese hastening the downfall of the Ming; in 1894, the defeat of the Qing Dynasty over the status of Korea spelled the end of China's system of tributary relations. This defeat set the stage for Russo-Japanese rivalry in Northeast China (Manchuria) that led not only to the Russo-Japanese War in 1905, but set in motion the process of creeping annexation that nearly ended Chinese sovereignty and led directly to the Pacific War in the 1940s. The People's Republic of China's (PRC) entry into the Korean War in October 1950 marked the true beginning of China's achievement of great power status. China's interest in Korea therefore transcends Cold War politics and ideology. The security of the Korean frontier is crucial to China's own security. Moreover, Sino-Korean relations can never be disentangled from the Sino-Japanese relationship.

China has demonstrated a consistent pattern of loyalty to its client. This was as true at the dawn of modernity in 1592 as it was in 1950, when it challenged the world's leading power even without Soviet air cover after UN troops under the command of the U.S. crossed the 38th Parallel into the territory of North Korea. The pattern of client loyalty is complicated today by China's extensive and close relationship with the Republic of Korea (ROK or South Korea). However, PRC commitment to the Democratic People's Republic of Korea (DPRK or North Korea) is as much a product of longstanding principles of regional international behavior as it is of ideology and history.

Chinese interests in the Korean peninsula can be stated in the following order:

1. Stability;
2. Demonstrating loyalty as a patron;
3. Maintaining a geostrategic buffer between itself and the U.S. and Japan;
4. Enhancing its status through great power diplomacy.

Stability places limits on the role of external powers on China's borders; instability invites intervention. Loyalty enhances China's reputation as a great power patron in the region – and trumps Russia in Pyongyang. Maintaining an allied DPRK regime (i) provides a buffer against U.S. (ground) forces; (ii) keeps refugees out of China's Northeast; (iii) enhances the social and economic stability of the ROK with which China maintains close economic ties; (iv) provides China with a potentially privileged economic hinterland. Playing a privileged broker role provides China with leverage in its relations with the U.S. and outmanoeuvers Japan on the Northeast Asian mainland.

The North Korean nuclear test of October 9, 2006 was a severe shock to China. In the days between Pyongyang's declaration of its intention and the actual test on October 9 China explicitly called on North Korea to show restraint.[1] The test came at a particularly sensitive time for China's leaders, coinciding with the Sixth Plenary Meeting of the Chinese Communist Party Central Committee, at which a major resolution on Hu Jintao's domestic program of social reform, aimed at building a "harmonious society."[2] It characterized the test as "flagrant" *hanran* and unacceptable.[3] It was also a considerable loss of face.

For three years China had hosted the six-party talks in Beijing in an effort to stabilize the Korean peninsula and restore its status as nuclear free. In the effort to dissuade the DPRK from pursuing the option of becoming a declared nuclear power and basing its security on nuclear deterrence, China, in parallel to the six-party talks, sought to engage the North Korean regime through enticing it into a policy of market opening and direct investment from China. This policy was closely coordinated with similar policies undertaken as part of South Korea's "Sunshine Policy." China's overall strategy was to induce an atmosphere of "win-win" or a positive sum game where Pyongyang engagement with the international community, including the United States, would increase the security of its regime and open up horizons for improved economic prosperity. Far from simply "propping up" the Kim family regime, China's leaders went to considerable lengths to engage the U.S. and patiently guide Pyongyang back to the negotiation table. China not only worked hard to prepare a formal declaration of principles, but persuaded Washington to refer to policy that it had implicitly and explicitly discarded—that of the Framework Agreement of 1994. In comparison with 1994 when China refused to expend any political capital in pursuit of a regional agreement, despite urgent requests both from Seoul and Tokyo, China's investment of political capital in the current context has been substantial, material, and persistent. The proper question with respect to China's role in the Korean nuclear issue is not why so little, but this time, why so much.

---

1 Zhongfang xiwang Chaozai heshiyanwentishang baochi lengjing he kezhi [China hopes that North Korea will demonstrate coolness and self restraint with respect to the question of nuclear tests] online at <http://world.people.com.cn/GB/1029/42354/4884815.html> (Accessed October 5, 2006).

2 <http://news.xinhuanet.com/english/2006-10/11/content_5190592.htm> (Accessed November 7, 2006)

3 Waijiaobu fayanren jiu Chaoxian he shiyan wenti da jizhe wen [Foreign Ministry Spokesperson answers journalist questions about the North korean nuclear test] online at

<http://www.pzh.gov.cn/Article/gjzx/gnei/200610/6389.html> (accessed October 11, 2006)

Some might argue that China has done so out of realist concerns for its own security and out of its own fears of a deteriorating security dilemma with its neighbors, especially Japan, should North Korea exercise the nuclear option. However, this was essentially the same prospect that China faced in 1994 when no similar Chinese initiative was forthcoming. What has changed?

While we can point to the growing intimacy of China with the Republic of Korea both economically and politically as well as the corresponding but largely unconnected deterioration of the Sino-Japanese relationship, I would argue that the single most important factor has been China's changing views of its international role, and in particular, its turn towards multilateral engagement that became increasingly obvious at the beginning of the new millennium. In short, China has sought through the six-party talks to put into practice its "new security concept" and assume the role of a "responsible great power." This effort has coincided especially with the emergence of Hu Jintao at the helm of the Communist Party of China, with his declared policy of seeking a "harmonious society" domestically and a "harmonious world" internationally. Underpinning China's desire to engage multilaterally on this issue has been a recognition that China's security is not a simple matter of choosing friends and confronting enemies. For example, China has been wary of U.S. efforts to develop a missile shield in the Asia- Pacific particularly its cooperation with Japan over theatre missiles defense (TMD).[4] This was seen as a threat both to Chinese efforts to engage in coercive diplomacy over Taiwan independence and possibly to China's nuclear deterrent. Following North Korea's missile tests in 1999, China's then Premier Zhu Rongji publicly asked: "Another pretext for the deployment of TMD by Japan and the U.S. is that the DPRK has launched missiles and is developing nuclear weapons while China has failed to exercise its influence. We have no knowledge of these activities. Notwithstanding the fact that [the DPRK] is an independent how can we interfere?"[5]

By the late 1990's China had begun to herald a "new security concept." In 1999 China's Chief of Staff Zhang Wannian wrote "History proves that simply by increasing military preparedness and military alliances will not gain long-term peace. We can create real peace only by linking together mutual trust and common interests through dialogue and cooperation."[6] This idea was enshrined in China's 2002 Defence White Paper: "To enhance mutual trust through dialogue, to promote common security through cooperation, and to cultivate a new security concept featuring mutual trust, mutual benefit, equality and benefit have become the requirements of the new era."[7] The new thinking recognizes that "security is indivisible" and that one cannot enhance one's own security without taking account of the security of one's neighbours and potential adversaries. This concept was enshrined in Jiang Zemin's speech to the

---

4 On this issue see Zhu Feng, *Dandao daodan fangyu jihua yu guoji anquan* [Ballistic Missile defense and international security] (Shanghai: Renmin Chubanshe, 2001).

5 Press conference of the Second Session of the 9th NPC RMRB March 16, 1999.

6 Zhang Wannian, *Dangdai Shijie junshi yu Zhongguo guofang* [Contemporary World Military affairs and China's Defence] (Beijing: Junshi kexue chubanshe, 1999) p. 176.

7 "China's National Defense in 2002" online at <http://english.people.com.cn/features/ndpaper2002/ndl.html> (accessed December 9, 2002).

Sixteenth Communist Party Congress in 2002, "In the area of security, countries should trust one another and work together to maintain security, foster a new security concept featuring mutual trust, mutual benefit, equality and coordination, and settle their disputes through dialogue and cooperation and should not resort to the use or threat of force."[8]

For China, and for Hu Jintao personally, the Korean nuclear crisis that broke out in October 2002 has been a test case of solving security problems through dialogue and through seeking "win-win" solutions. The Korean nuclear issue poses a particularly difficult dilemma for the Chinese leadership, since it pits a traditional ally and friend its "lips and teeth" relationship with the DPRK with one of its largest trade and investment partners and a key partner in forging a new regional identity in East Asia, the Republic of Korea. Moreover, the priority foreign policy objective for China's leaders throughout the reform era has been to stabilize relations with the world's leading and now sole superpower, the U.S. China could not remain aloof from the crisis this time without risking a deterioration in the Sino-American relationship that was key to its goal of "peaceful emergence." The fact that in 2003 the U.S. directly solicited China's help in defusing the crisis multilaterally in contrast with its initially bilateral solutions in 1994 and 1998 is testimony to the relativization of U.S. power and influence in Northeast Asia, if not its actual deterioration, due not only to the rise of China but also to U.S. preoccupation with Iraq and the Middle East. This relativization of U.S. power and influence was a potential platform of opportunity for China to assert a positive regional and global role, a role that China took on assiduously so long as it could be kept within the parameters of its own perceptions of its proper role and identity. That is, China was never prepared to go beyond its vision of itself as benevolent great power that sought peaceful solutions through dialogue and positive sum strategies.

The current North Korean nuclear crisis thus occurred at a time of great dynamism in Chinese diplomacy, a period of rising confidence as well as much greater initiative in the area of multilateral relations. In fact while official Chinese diplomacy on the nuclear issue was conciliatory, cautious and surprisingly even-handed (given the DPRK's status as the PRC's ally), the range of views addressed with the Chinese policy community by academics and foreign policy observers has been wide and contested. At one extreme, there have been voices calling for China to drop its alliance with North Korea in favor of regime change and strengthening ties with South Korea;[9] at the other end of the spectrum are those who adopt a "realist"

8 Jiang Zemin, "Speech to the 16th National Congress of the Communist Party of China," online at <http://news.xinhuanet.com/english/2002-11/18/content_632571.htm>.

9 Prominent among these have been People's University's Shi Yinhong, who wrote in February 2003 that China might be subjected to North Korea's nuclear blackmail in Hong Kong's Dagong Bao see, China Asserts It has Worked to end Korea's Nuclear crisis, *New York Times*, February 13, 2003 online at <http://www.nytimes.com/2003/02/13/international/asia/13BEIJ.html?ex=1046137979&ei=1&en=b6102b744eec1b57>;TianjinAcademy of Social Sciences Wang Zhongwen "Guancha Chaxian he dongbei Ya qingkuang de xin guandian" [A new viewpoint to observe the North Korea and the situationin Northeast Asia] *Zhanlue yu Guanli,* no. 4 (July/August 2004) ; and Peking University's Zhu Feng. For Zhu Feng's views see "Zhongguo Chao he zhengce he celue de bianhua," *Zhongguo Zhanlue*, no. 3 (2004), pp.

stance that can even embrace nuclear proliferation as a hedge against U.S. support of Taiwan.[10] Underlying these contradictory approaches have been two opposed outlooks on China's foreign policy outlook. The first sees China as effectively aligned with the liberal world in pursuit of free trade and capitalist development, while the other sees China locked into a long term struggle with the U.S. over U.S. efforts at global hegemony and regional containment of China's rise. Official Chinese policy has carefully avoided encouraging either of these two extremes, preferring to believe that it was possible to enhance the status quo by encouraging the U.S. to guarantee North Korea's security while encouraging Pyongyang in turn to exit its economic crisis by opening up its economy. Beijing saw the future through the integration of the DPRK in the international community, a goal that it saw as the interest of all parties.

It did not abandon this strategy even after it suffered a loss of face. China first urged Pyongyang not to go ahead with its missile tests in July 2006, and was dismayed when these were carried out. Right at the time when the DPRK was planning its missile tests, China was preparing to send a high level delegation to China to celebrate the 45th Anniversary of the PRC-DPRK Friendship Treaty.[11] While China viewed the missile tests as provocative and unhelpful,[12] Hu Jintao nevertheless decided to go ahead with Politburo Member Hui Liangyu's visit. During the visit Kim Jong-Il pointedly refused to see Hui, relegating him to a protocol visit with the Premier.[13] During the visit it appears the two sides disagreed over the nature of the 1961 Treaty. China pointedly referred to its role in friendship and development and made no mention of its status as an alliance. However the text of the 1961 Treaty makes this fact abundantly clear, as the text of Article II makes it clear:

1-9. The Academy of Social Sciences Shen Jiru also warned that a military crisis on the Korean Peninsula could cut China's GDP growth by 10-20 per cent and therefore advocated scrapping and openly renegotiating the alliance provisions of the PRC-DPRK Friendship, Cooperation and Mutual Support Treaty of 1961. See, Shen Jiru, "Weihu Dongbei Ya anquan dangwu zhi ji" [The Current urgency of maintaining security in Northeast Asia] *Shijie Zehngezhi yu Jingji*, no. 9. (2003), pp. 1-7.

10 The most prominent among these is Fudan University's Shen Dingli, who first pronounced his views in "Accepting a Nuclear North Korea," vol 168, no. 3 (March 2005); he extends these views in "How To Get to be a de Facto Nuclear Power," *Le Monde Diplomatique*, November 2006. Yan Xuetong is less sanguine about North Korea's ambitions, but Recognizes that Pyongyang acts out of concern for its regime's security. Author's interview in December 2006.

11 "Chinese Vice-Premier to Visit North Korea," online at <http://politics.people.com.cn/GB/1024/4581584.html> (Accessed December 21, 2006).

12 "Foreign Ministry spokesperson reacts to North Korean Missile Test" online at <http://www.fmprc.gov.cn/ce/ceae/chn/wjbfyrth/t261535.htm>; See also, Dagong Bao Zhongguo fandui huoshang jia you [China Opposes adding fire to the flames] online at <http://www.chinadaily.com.cn/hqtt/2006-07/12/content_639271.htm>.

13 « Hui Liangyu huijian Chaxian Neige Zongli Bu Fengzhu » [Hui Liangyu Sees North Korean Premier Park] online at <http://politics.people.com.cn/GB/1024/4581584.html> (accessed January 6, 2007).

> The Contracting Parties undertake jointly to adopt all measures to prevent aggression against either of the Contracting Parties by any state. In the event of one of the Contracting Parties being subjected to the armed attack by any state or several states jointly and thus being involved in a state of war, the other Contracting Party shall immediately render military and other assistance by all means at its disposal.[14]

While in the DPRK capital Vice-Premier Hui apparently urged the ROK and U.S. not to conduct upcoming military exercises. He also apparently condemned the draft resolution that Japan circulated at the UNSC as "excessive." However, these references were later expunged from Chinese websites.[15] This appears to set the stage for UN Security Council Resolution 1695, passed on July 15, supported by China, which condemned the DPRK missile tests.[16] By early August China's Foreign Ministry spokesperson Liu Jianchao denied that China had any influence over North Korea and insisted that they were two sovereign states that decided on their foreign policy on their own. Given the deteriorated state of relations before the beginning of September 2006, the reports that China cut off oil deliveries are credible. This is also indirectly corroborated by the retaliatory act by Pyongyang in early September to cut off Chinese tourism to Pyongyang.[17]

The nuclear test of October 9, 2006 was in direct contravention of the North Korean pledges to denuclearize the Korean peninsula that had been included in every joint communiqué of every meeting between Chinese and DPRK leaders since the 1990s. China was therefore forced to react.[18] During the Press Conference that followed China explicitly repudiated that it had a military alliance with any country including the DPRK. It reluctantly joined in voting sanctions against the DPRK in supporting at the UN Resolution 1718 at the Security Council.

---

14 Sino-Korean Treaty of Friendship, Cooperation and Mutual Assistance, July 11, 1961 text as appears in *Peking Review*, vol. 4, no. 28, p. 5 available online at <http://www.marxists.org/subject/china/documents/china_dprk.htm>.

15 The Chinese report initially carried Hui's urging during his meeting with Premier Park. but subsequent reports merely referred to North Korean warnings issued at a DPRK Cabinet meeting. See <http://world.people.com.cn/GB/1029/42354/4586029.html> (Accessed December 21, 2006).

16 UNSC Resolution 1695 (2006) S/res/1695(2006).

17 "North Korea Floods Stop Chinese tourists, agents say " Reuters September 2, 2006 available online at

<http://www.hkhkhk.com/engpro/messages/2155.html>.

18 "Question: China is North Korea's Close Neighbour and close ally, and is also a nuclear power. We can understand if China hopes that the Korean nuclear test is a secure nuclear test. Did the Chinese provide any technical help to ensure that the North Korean nuclear test was safe? "I do not agree with your statement just now that China is North Korea's ally. China has a policy of non-alignment China's relations with North Korea are established on the basis of the principles of international relations" online at <http://news.enorth.com.cn/system/2006/10/10/001429913.shtml>.

The History of the Six-Party Talks

While China had originally remained aloof in 1994 by the late 1990s China was expressing support for the Agreed Framework and responded positively to the U.S. proposal to convene four-party peace talks to replace the Korean Armistice with a permanent peace treaty. These talks took place at Columbia University in New York on August 5, 1997.[19] These talks failed to reach an agenda after two rounds of meetings in August and September 1997 due to Pyongyang's insistence that the ROK-U.S. alliance be made a subject of the talks. China remained engaged on the Korean issue during the missile crisis that broke out following Pyongyang's sudden testing of the Taepodong I ballistic missile over the Japan Sea on August 1998. However, President Clinton's clear commitment to engage Pyongyang towards the end of his presidency left little room for an independent Chinese initiative. By contrast, it was the refusal of U.S. President George W. Bush to engage North Korea following his election in 2000 that provided space for Chinese initiatives.

Following the outbreak of the current nuclear standoff as a result of U.S. Undersecretary of State Kelly's meetings with his North Korean counterpart in October 2002, the U.S. responded by urging efforts at a multilateral diplomatic solution. China responded as early as April 2003. After attending the inauguration of South Korean President Roh Mu-hyun in Seoul, Chinese Vice premier Qian Qichen, in one of his last diplomatic activities before retirement, attempted to convene the U.S. and North Korean sides in Beijing. When these talks failed to yield results, China redoubled its efforts. It appears that Pyongyang preferred a scenario which allowed it to engage bilaterally with Washington, while the Bush administration insisted on a wider multilateral forum. In July 2003, Chinese President and Party General Secretary Hu Jintao wrote to Kim Jong-Il.[20] Hu Jintao promised greater economic aid while reassuring Kim that China would work to achieve a security guarantee for the DPRK in return for denuclearization of the Korean peninsula. The letter was delivered by Chinese Vice-Foreign Minister Dai Bingguo in Pyongyang. Dai was the former head of the Chinese Communist Party's International Liaison Department, and had personally accompanied Kim Jong-Il on his visits to China in May 2000 and January 2001. His efforts led to the convening of the six-party talks in Beijing in August 2003. China summoned both Koreas, Japan and Russia, in addition to the U.S. to hammer out an agreement to denuclearize the Korean Peninsula.

Talks dragged out for some two years without visible resolution, during which North Korea's nuclear program has continued to move forward, reprocessing plutonium into what is estimated at 4-6 nuclear warheads. The talks bogged down over U.S. insistence that Pyongyang return without preconditions to the status quo ante under the terms of the Agreed Framework and the NPT by thoroughly dismantling its nuclear program in a permanent, transparent and verifiable manner. A multilateral setting let Washington off the hook of dealing with a loathed regime directly, but also

---

19 <http://www.cnn.com/WORLD/9708/05/korea.talks.pm/> (Accessed October 31, 2006).

20 Hong Kong Economic Journal, August, 28, 2003. online at <http://www.nautilus.org/archives/pub/ftp/napsnet/special_reports/HuJinTao-Letter.txt> (November 30, 2006).

held out hope that China might subtly pressure its wayward ideological kin into line. China, however, was against painting Kim Jong-Il into a corner that might provoke his reckless streak further. Besides, both China and South Korea feared the impact of an imploding North Korean regime on their own population and resources. Meanwhile South Korea continued its "sunshine policy," offering economic inducements to their northern brethren over Washington's objections. Happy to play host, China also urged the U.S. to do more bilaterally.

Pyongyang set a high price for its agreement. This included an ironclad guarantee of its sovereignty and the security of its regime against talk of "regime change" emanating out of Washington, as well as tangible economic benefits for its energy and food-starved totalitarian dystopia. China's initial role was mainly to provide the hexagonal table and to keep the parties engaged. As the talks continued, they revealed a pattern that saw China and South Korea closely aligned on the need to keep North Korea engaged and to resolve the issue peacefully, while the U.S. and Japan took a harder line. China failed to hammer out a joint statement promising to peacefully resolve the nuclear issue in the second round of talks in February 2004, having to settle for a statement from the Chair. No further progress was reported after the third round, although here Pyongyang first signaled its willingness to accept an agreement to be implemented in phases with tangible benefits resulting from each step. One indicator of the importance and price China was willing to pay in order to get these talks underway was the decision to shut down the influential Beijing journal *Strategy and Management* [Zhanlue yu Guanli] which in August 2004 carried an article by Tianjin researcher Wang Zhongwen, accusing the North Korean regime of starving its own people and damaging China's ties to the U.S. by its policy of brinkmanship.[21] Interview subjects made it clear that Pyongyang insisted that the article be withdrawn and the journal disciplined if it was to return to the six-party talks.[22]

As the talks bogged down and Pyongyang began to threaten to test a nuclear device and declare itself a nuclear weapons state, South Korea and China began to coordinate their policies closely. The ROK insisted that the fourth round include a formal statement, but also offered an energy package as an inducement. In the spring of 2005 each side made moves that enabled a breakthrough. Washington expressed its lack of aggressive intent towards Pyongyang. Pyongyang, all the while insisting on its sovereign right to possess nuclear weapons, signaled its willingness to enter an agreement to implement a nuclear-free Korean peninsula in stages. Seoul offered the North vast amounts of free electricity flowing across the DMZ in exchange for abandoning the nuclear option. The U.S. and the DPRK resumed bilateral contacts through the medium of Pyongyang's UN ambassador, who also traveled to Canada, to telegraph eagerness for an accord. When the fourth round of the six-party talks reconvened in July 2005, the envoys of Washington and Pyongyang envoys met directly on the fringes of the hexagonal table.

21 "'Periodical in China Shut Down after article on North Korea" *The South China Morning Post,* September 22, 2004.

22 Author's Interview with Prof.Shi Yinhong, December 2004.

A Historic Agreement

On Monday, September 19, 2005 at the conclusion of the fourth round of talks regarding the nuclear crisis in the Korean peninsula, the six parties set out a joint statement.[23] The statement was cemented during the Mid-Autumn Festival cherished equally by Koreans and Chinese. China, with the active participation of the ROK, pushed for a formal written document to come out of the fourth round. The Chinese delegate acted as the crucial go-between to supply language that would satisfy both U.S. and DPRK requirements. China persuaded the U.S. to include language about potential supply of a Light Water Reactor, worked to beef up South Korean obligations to balance them against the obligations to be undertaken by the North, and worked to satisfy U.S. demands to put in step by step compliance on the basis of word for words action for action.

The most important contribution of China, was to broaden the framework of the agreement beyond the issue of denuclearization to include comprehensive provisions for peace in the Korean peninsula, mutual recognition, and economic development. China effectively sought to put into practice the idea of "cooperative security" whereby the security of each party is contingent on the security of other member parties to the agreement. To this end, the security concerns of all sides would be taken into account in the final document. This flew in the face of the initial U.S. position that sought to limit the scope of the six-party talks to the issue of North Korea's denuclearization before any ancillary issues could be entertained. Here, Pyongyang also for the first time, had to recognize the legitimacy of its southern sister regime in Seoul in a framework document signed onto by other parties. Pyongyang had always maneuvered to deal bilaterally with the South or with other parties separately (U.S. or Japan) while excluding the South, thereby preserving the myth that it was the only legitimate spokesperson for Korean nationalism. Now it had to recognize the legitimacy of the Seoul government in its relations with other great powers.

The September 19 Joint Statement has six clauses (compared with four clauses in the 1994 Agreed Framework). The first clause reaffirms the goal of denuclearizing the Korean peninsula but reserves for Pyongyang the right to peacefully develop nuclear energy. The second reaffirms mutual respect for the sovereignty of the participant states, and sets the normalization of diplomatic relations between Pyongyang and Washington and Tokyo respectively as an explicit objective. Clause 3 calls for improved trade and economic relations and refers directly to Seoul's promise to supply the North with electricity. Clause 4 calls for a peace treaty to permanently settle the Korean War; Clause 5 agrees on the principle of achieving agreement through step by step reciprocal actions and the last clause refers to the current (fifth) round of talks in November.

Overall, the agreement marks a milestone in establishing China's strategic importance as a diplomatic power in Northeast Asia. China has now helped broker

---

23 BBC News – Sept 19: Full Text of the 4th Round Declaration of the 6-Party Talks online at <http://news.bbc.co.uk/1/hi/world/asia-pacific/4259176.stm> (Accessed Sept 28, 2005).

an agreement that the world's sole superpower was unable to make on its own. The agreement makes a grand bargain between denuclearization and Pyongyang's desire for international recognition and material aid. Not only does it tie nuclear concerns with DPRK's desire for international recognition, for the first time it commits the DPRK and the ROK formally to each other in a treaty context countersigned by other powers. In effect, China has posed itself as the guarantor of the status quo in Northeast Asia. This recognizes the legitimacy of two sovereign states on the Korean peninsula and links this with the need to normalize U.S.-DPRK and Japan-DPRK relations. It is a draft peace treaty by any other name, which at long last moves Northeast Asia beyond the Cold War.

Despite China's concerted efforts to put in place an agreement that satisfied the material and security interests of all sides, the agreement threatened to unravel almost before the ink was dry. Earlier in September 2005, the U.S. Treasury Department had put the DPRK on notice with regard to the activities of Banco Delta Asia (BDA) in Macau with respect to money laundering and trafficking in counterfeit U.S. currency. When immediately after the signing of the September 19 Agreement Pyongyang began to demand a Light Water Reactor (as though the 1994 Framework Agreement had been restated without pre-conditions). The U.S. then reacted by placing financial sanctions on BDA Macau.[24]

China Presses Ahead

China nevertheless signaled the importance of the September 19 Statement by scheduling a visit by President Hu Jintao to Pyongyang within a month of the statement's release. This was the first visit of China's Head of State, Party Chief and Commander in Chief since Jiang Zemin last visited in 1998. Hu pointedly visited both the Taian Glass Factory recently constructed with Chinese Aid, and the Korean countryside to show China's concern with economic development in North Korea. Hu Jintao's speech at the state banquet on October 28 pointedly emphasized China's achievements in economic growth since opening its economy to the world in 1978.[25] In his press conference, the Head of the Chinese Communist Party's International Liaison Department, Wang Jirui, who accompanied Hu on his visit, stressed that this visit had broadened the framework of China's relationship with the DPRK beyond the past pattern of fraternal parties and bilateral military alliance towards economic cooperation within the framework of the globalizing political economy.[26] Getting Kim Jong-il to acknowledge the success of China's development model by highlighting China's contribution to North Korea's economic development, was China's effort to make North Korea's linkages with global market economies irreversible. China's intention is to bring the DPRK into international society and the global political

---

24 See Anthony DiFilippo, "Hubris Intransigence and the North Korean Nuclear Crisis" in *Japan Focus* January 8, 2007 online at <http://japanfocus.org/products/details/2316>.

25 <http://news.xinhuanet.com/world/2005-10/29/content_3698246.htm> (Accessed October 28, 2006).

26 <http://news.xinhuanet.com/world/2005-10/30/content_3703660.htm> (Accessed October 30, 2006).

economy. By acknowledging China's role in North Korea's own economy, Kim has to jettison part of his father's *Juche* legacy. In altering the calculus of Kim Jong-Il by loading the dice in favour of global engagement China steers Pyongyang away from the temptation to behave as a "rogue" state. A Chinese commentator characterized Hu's visit to Pyongyang as an effort to bring the relationship between the PRC and the DPRK into the ambit of normal relations among states, away from a Cold War relationship based on ideology and geopolitics.[27]

China went further towards this goal by hosting Kim Jong-Il in China early in the new year. From January 10-18, 2006 China showed Kim the benefits of open trade through a tour of the Special Economic Zones of Zhuhai and Shenzhen in the Pearl River Delta region of Guangdong Province, adjacent to Hong Kong, as well as the Three Gorges Hydroelectric power station and an agricultural research station along the Yangtse River province of Hubei. The priority of the tour was to demonstrate the benefits of an open economy as well as to show off China's capacity in areas that are critical for the DPRK, such as of electric energy and agriculture, electronics and telecommunications. The key priority for China was to get the DPRK to change its development model. The official report of the visit recorded Kim's explicit endorsement of China's open policy development model.[28] The importance given to this visit by the Chinese leadership can be gauged by the fact that during Kim's visit he was accompanied or met will all of the members of the Standing Committee of the Politburo of the Chinese Communist Party and 16 out of 25 members of the wider Politburo. China is clearly offering Kim extraordinary political solidarity premised on Kim's acceptance of a foreign policy of international interdependence.

At the beginning of 2006 North Korea refused to return to the six-party talks, citing "provocations" and the "hostile attitude" of the U.S. namely, the imposition of financial sanctions on North Korea's banking. Despite this, China persisted in calling on Pyongyang to return to Beijing. Once again, after North Korea refused to heed China's cautions over missile tests in July 2006 Pyongyang went ahead anyway. It is now clear that Pyongyang's repeated flouting of China's patient gestures resulted in a cutoff of China's oil exports in September, alongside the imposition of financial sanctions.[29] This may have been one reason why North Korea suddenly cut off Chinese tourism in September. Some sources even claim that top DPRK military figures insisted on carrying out the nuclear test to show Beijing that Pyongyang was not its client.

---

27 3 "Interpreting a new form of Friendship" online at <http://news.xinhuanet.com/world/2005-11/02/content_3717245.htm> (Accessed November 2, 2006).

28 Xinhua, January 18, 2006 online at <http://english.sina.com/china/p/1/2006/0122/62949.html> (accessed March 3, 2006).

29 "Behind China's Stance on North Korea," *The Wall Street Journal*, November 6, 2006 online at <http://online.wsj.com/article/SB116277002861613964.html> (Accessed November 5, 2006). See also, "China May be using Oil to Press North Korea," *New York Times*, October 31, 2006.

China' Two Pronged Approach to North Korea's Provocation

This set the stage for China's extraordinary public display of disapproval following the nuclear test of October 9. Even then, China continued to call for a level-headed response and diplomacy.[30] China did not make any public display of the sanctions it was applying behind the scenes. China toned down the punitive rhetoric and opposed sanctions that would cut off North Korea's trade outright. Hu Jintao spoke to U.S. President Bush on October 10 and China responded by voting for UN Security Council resolution 1718 on October 14, 2006. This was followed up by State Councilor Tang Jiaxuan's visit to Pyongyang October 19. China pledged to uphold the UN sanctions but did not stop legal trade with the DPRK.[31] Tang was the personal emissary of President Hu Jintao and took with him China's chief negotiator to the six-party talks, Vice-Foreign Minister Wu Dawei as well as the director of the CCP Central Committee's Liaison Department. As a result Pyongyang pledged not to go ahead with a second nuclear test.[32] China insisted on normal trade relations alongside the sanctions that targeted weapons trade and the supply of luxuries to Kim's dinner table. China kept dangling the carrots, while making clear that it was prepared to let Kim sit a time out. In the meantime, Beijing played host successively to the incoming Japanese Prime Minister Shinzo Abe, the South Korean Foreign Minister, and incoming UN Secretary General, Ban Ki-moon, as well as U.S. Secretary of State Condoleeza Rice. Subsequently, U.S. chief negotiator to the six-party talks, Undersecretary of State Chris Hill returned to Beijing. On October 31 China engaged North Korea and U.S. Chief negotiators in direct talks in Beijing as a result of which China achieved agreement that the six-party talks would resume before the end of the year (2006).[33]

On November 28 2006, China's Chief negotiator Wu Dawei met with his U.S. counterpart Undersecretary of State Chris Hill and his North Korean opposite number. While Pyongyang sought to gain recognition as a nuclear power, the Chinese have insisted on the denuclearization of the Korean Peninsula as the direct aim of the six-party talks.[34] China's public stance is therefore more forthcoming than the position of some academics, especially Shen Dingli of Fudan University who has suggested that

---

30 Waijiaobu fayanren jiu Chaoxian heshiyan Anlihui jueyi wenti fabiao tanhua [The Foreign Ministry spokesperson publishes speech on the North Korean Nuclear Test and the UN resolution] online at <http://www.pzh.gov.cn/Article/gjzx/gnei/200610/6803.html> (Accessed October 16, 2006).

31 See "Sanctions don't dent China-North Korea Trade," *New York Times*, October 27, 2006.

32 "Chinese envoys arrive in Pyongyang," <http://today.reuters.com/news/articlenews.aspx?type=politicsNews&storyID=2006-10-19T152819Z_01_N14268552_RTRUKOC_0_U.S.-KOREA-NORTH.xml&WTmodLoc=PolNewsHome_C2_politicsNews-4> (accessed October 20, 2006); <http://www.chinanews.com.cn/gj/news/2006/10-24/809268.shtml> (Accessed November 7).

33 <http://www.chinanews.com.cn//gn/news/2006/11-02/814484.shtml> (Accessed November 7, 2006).

34 See the remarks of Foreign Ministry Spokesperson Liu Jianchao November 28, 2006 online at <http://world.people.com.cn/GB/1029/5100668.html>.

China may see some degree of proliferation as in its interest as leverage over the U.S. in its pursuit of concessions over Taiwan.[35] It is in China's interest to consolidate its relationship with South Korea and stabilize its relations with Japan. For this reason, and because of a hard headed calculation of North Korean dependence on China, it is unlikely to concede to North Korean demands concerning nuclear power status. At the same time, neither will China pressure North Korea towards regime collapse.

The Fifth Round

In early December 2006 a breakthrough was made which allowed the long-stalled fifth round of the six-party talks to resume. The U.S. offered to discuss the financial sanctions in parallel to resumption of talks about the implementation of the September 19, Joint Statement, and reportedly was willing to make a far reaching economic proposal.[36] Accordingly, talks resumed in Beijing December 18, 2006. The U.S. put forward a detailed timetable for the implementation of the Joint Statement according to the principle of "step-by-step" and "action for action." China was both pleased by its efforts to press the DPRK and U.S. sides to engage bilaterally as well as pleased by the U.S. proposal to get the talks moving.[37] Initially the U.S. side shared China's optimism.[38] Ultimately, the talks again adjourned without agreement, according to the U.S. side because Pyongyang's negotiators had no mandate to reach an agreement, and had a non-negotiable mandate for the removal of the financial sanctions.[39]

Nevertheless, the fact that the Chinese chose to publish the details of the U.S. offer with favorable comment demonstrates that they were generally pleased by the U.S. offer and the U.S. approach.[40] It is not at all surprising that the visible shift of the Chinese side towards the posture adopted by the U.S. has also been accompanied by a considerable hardening of the Chinese position with respect to the nuclear ambitions of Iran as evidenced by its support for UNSC 1737 (December 27, 2006) concerning

35 See Shen Dingli, "Implications of the DPRK Nuclear Test," *Le Monde Diplomatique,* October 23, 2006.

36 "U.S. Offers North Korea Aid for Dropping Nuclear Plans," *New York Times,* December 6, 2006 online at <http://www.nytimes.com/2006/12/06/world/asia/06korea.html> (Accessed December 6, 2006).

37 See the Statement by host Wu Dawei "Liufang huitan xiuhui Zhongfan fa zhuxi shengming" [The six-party Talks adjourn, the Chinese side issues Chairman's declaration] online at <http://world.people.com.cn/GB/1029/5204520.html>.

38 "U.S. Sees Progress in Talks With North Korea," *New York Times*, December 20, 2006 online at <http://www.nytimes.com/2006/12/20/world/asia/20cnd-korea.html?hp&ex=116667720 0&en=250e76fdf345be3b&ei=5094&partner=homepage>.

39 "North Korea Nuke Talks End Without Deal," *Washington Post*, December 22, 2006 online at <http://www.washingtonpost.com/wp-dyn/content/article/2006/12/22/AR2006122200242.html>.

40 Tao Wenzhao, "Still hope to settle DPRK nuclear issue," <http://www.chinadaily.com.cn/opinion/2006-12/26/content_767483.htm> (Accessed December 26, 2006).

Iran's nuclear enrichment activities.[41] In both cases, China has qualified the stance of the DPRK and the Islamic Republic of Iran as defiance of the "international community."[42] The key to understanding China's current stance is not only that it favors multilateral solutions over unilateral and bilateral ones, but that it is no longer afraid to modify and even shift longstanding bilateral relationships in support of a multilateral diplomatic solution. In the case of the DPRK, China has effectively abandoned its mutual defense obligations in favor of reaching a multilateral solution that promises two things: stabilization of Sino-American relations and the creation of a permanent framework for Northeast Asian security where China holds down a secure corner of a multilateral framework.

Conclusion

From interviews and from creative interpretation of China's approach at the six-party talks it is clear that China does not expect that even in the most optimistic scenario, the DPRK would give up its nuclear weapons except at the end of a long process of confidence building, security assurance, and economic aid and development. In other words, China expects to see the six-party talks evolve into some kind of framework akin to the progressive stages of arms control engaged in by the Soviet Union and the United States (SALT and START) in the final phases of the Cold War. Further, it seems also clear that that China sees the Joint Statement as a skeleton or draft peace treaty which would supersede China's bilateral mutual defense treaty with the DPRK. As one commentator puts it, "China engages positively in the six-party talks because there is virtually no restriction on China, while China can play a guiding role in this [forum]."[43] He sees four reasons for China to engage multilaterally: (i) to avoid being pulled into a war and to maintain regional stability; (ii) to avoid nuclear proliferation and a regional arms race; (iii) to improve China's image as a "responsible great power"; and (iv) to enhance China's role as a stakeholder on the Korean peninsula.[44] It is clear that with very few exceptions, both the Chinese government and the Chinese foreign policy community see multilateral engagement over the nuclear issue as a "win-win" issue, where China's regional and global status can only be enhanced while improving its leverage both with respect to the U.S. and Japan.

The pattern of Chinese engagement has moved from low expenditure of political capital to increasingly higher expenditure of political capital. This capital is both symbolic and material. Some Chinese commentators fear that as a result

41 S/res/1737 (2006).

42 "Patience key to resolving nuclear issues," *China Daily*, November 20, 2006 online at <http://www.chinadaily.com.cn/world/2006-11/20/content_737134.htm> . See also MK Bhadrakumar, "China's Tortuous Middle-East Journey: Israel, Iran and Saudi Arabia," *Japan Focus*, January 15, 2007 online at <http://japanfocus.org/products/details/2322>.

43 Jiang Zhaijiu "Zhongguo dichu duobiananquan hezuo de dongyin," [The motivations for China's regional multilateral security cooperation] *Guoji Zhengzhi Kexue*, no. 1 (2006), p. 20.

44 Ibid, pp. 20-21.

of China's increasing tilt towards the U.S., China's leverage over Pyongyang has decreased.[45]On the other hand when one takes into account that China was already supplying more that 70 per cent of North Korea's oil requirements and 1/3 its food and other commodities virtually without compensation *prior* to the outbreak of the current nuclear crisis,[46] then China was already reaping a very poor return on its investment, both in terms of leverage over Pyongyang and in terms of global and regional prestige. China's efforts to return Pyongyang to the negotiation table have received widespread praise, not least from the U.S., a sign that China's global prestige has been enhanced by its role in the six-party talks.[47] While Pyongyang aspiration to nuclear deterrence has in some measure improved its security vis-à-vis the U.S. this unilateral effort to enhance security must be measured against the visible loss of the security guarantee provided by the 1961 PRC-DPRK Treaty of Friendship, Cooperation and Mutual Support. Today, while China continues to press the U.S. to provide explicit and written guarantees of non-aggression against the DPRK in return for denuclearization, China is itself only prepared to provide security as part of a multilateral arrangement that includes the U.S., the ROK, Japan, and Russia. This is not only a serious constraint on Pyongyang's diplomatic space but a shift in the way China manages regional security relations.[48] Through the six-party talks, China has not only managed to enhance its regional and global role as a diplomatic actor but has also significantly restructured the hub-and-spoke character of security relations in Northeast Asia centered on the bilateral security treaties between Washington, Tokyo and Seoul. Regardless of how the six-party talks eventually unfold, the Cold War structure of power relations in Northeast Asia has been transformed forever.

45 This position is taken by both Yan Xuetong and Shen Dingli (Author's interviews, December 2006).

46 See Shen Jiru, above. Also the official Chinese Foreign Ministry website entry on bilateral ties with the DPRK emphasizes the "free" economic aid China has provided the DPRK since 1994 see, DPRK- Bilateral relations available online at <http://www.fmprc.gov.cn/eng/wjb/zzjg/yzs/gjlb/2701/default.htm>.

47 See, for example "Diplomatic Victory for China," *The Independent*, November 1, 2006 online at <http://news.independent.co.uk/world/asia/article1945773.ece>. See also the author's "How China got North Korea Back to the Negotiating Table," *The Globe & Mail*, November 1, 2006.

48 See Ren Xiao, "Liu Fang huitan yu dongbei Ya duobian anquan jizhi de kenengxing" [The six-party Talks and the Possibility of a multilateral security system for Northeast Asia], *GuojiWenti Yanjiu*, no. 1 (2005), p. 41.

Chapter 6

Japan and the Second North Korean Nuclear Crisis

Yoshinori Kaseda

The end of the Cold War gave Japan and the Democratic People's Republic of Korea (DPRK or North Korea) a chance to significantly improve their diplomatic relations. They made some efforts for improvement, but their relations did not improve. Rather, they significantly deteriorated. The DPRK did not become a friendly neighbor to Japan. Instead, it became a major national security concern for Japan. In January 1991, Japan and the DPRK had the first round of their normalization talks. Yet, little progress was made, and talks broke down during the eighth round in November 1992, due partly to the two suspicions that the DPRK extracted plutonium from spent fuel rods and that the DPRK had abducted Japanese citizens.[1] Their bilateral relations further cooled down as a result of the DPRK's announcement of its intention to withdraw from the Nuclear Non-Proliferation Treaty (NPT) in March 1993 and its testing of a medium-range Nodong missile in May 1993, which demonstrated its capability to reach at least the southern half of Japan.

The international crisis stemming from this announcement was settled as a result of an agreement (Agreed Framework) reached in October 1994 between the United States and the DPRK. Japan played an important role in realizing their agreement by financing the construction of two light-water reactors (LWRs) in the DPRK. Yet, the tension between Japan and the DPRK built up again owing to the DPRK's launching of a Taepodong-1 missile in August 1998, which flew over Japan.[2] Later, tension eased as a result of Pyongyang's agreement to impose a moratorium on missile launches during its talk with the U.S. held in Berlin in September 1999. The settlement of the nuclear and missile issues, if not conclusive, led to the resumption of the normalization talks between Japan and the DPRK in January 2000 after seven years of suspension and to their first summit on September 17, 2002.[3] This summit

---

1 The strong suspicion originated from a revelation by a DPRK agent arrested for bombing a South Korean airplane in November 1987 that she learned the Japanese language and customs from a Japanese woman. The unknown Japanese woman was later identified by the Japanese police as a person missing since 1978.

2 The DPRK claimed that it was a rocket for sending a satellite into the orbit.

3 It was pointed out that Japan decided to hold the summit without a sufficient discussion with the U.S. and that the Bush administration did not like the idea. Toshimitsu Shigemura, *Gaiko Haiboku* (Tokyo: Kodansha, 2006), pp. 21-85; Masaru Sato and Haruki Wada, "Kitachosen to Abe gaiko," *Shukan Kinyobi*, 628 (October 27, 2006), p. 10.

produced a major breakthrough on the abduction issue as a result of the DPRK's admission of and apology for abducting thirteen Japanese citizens.[4] The DPRK also promised to continue its moratorium on missile launches beyond 2003. For its part, Japan gave the DPRK an official apology for the damage and suffering it caused to the people of Korea through its colonial rule and promised to provide North Korea with economic assistance, effectively as compensation, after diplomatic normalization, which was generally expected to be approximately 1 trillion yen.

However, North Korea's revelation that eight abducted Japanese citizens died in the DPRK escalated Japanese anger toward the DPRK, thus clouding the prospect for normalization. The prospect became even more uncertain after the U.S. revealed that North Korea was involved in uranium enrichment activity on October 16, just one day after the return of five abducted Japanese citizens from the DPRK. Pyongyang's missile tests in July 2006 and its nuclear test in October 2006 further aggravated its relations with Japan.

It is widely believed that North Korea has deployed approximately 200 Nodong missiles that can reach across Japan.[5] Therefore, it will generate a serious security concern for Japan if the DPRK improves its nuclear weapons technology and succeeds in developing nuclear warheads for those missiles.[6] Japan's national security is and will be most seriously affected by progress in the DPRK's nuclear weapons development. Japan thus has a very good reason to play an active role in stopping Pyongyang's attempt at developing operational nuclear weapons. However, as indicated by the deterioration of the problems associated with North Korea's nuclear development process, Japan has not been very successful in stopping North Korea's nuclear weapons development.

This chapter examines Japan's policy toward North Korea's nuclear issue. It focuses on Japan's response to the second nuclear crisis, which resulted from the U.S. revelation of Pyongyang's uranium enrichment activity in October 2002 and the subsequent collapse of the Agreed Framework of 1994. This study examines Japan's stance toward the DPRK at the six-party talks and outside the talks. It analyzes how the abduction issue has affected Japan's approach to the North Korean nuclear issue. It then discusses the factors that affected Japan's stance toward the nuclear issue and considers its future direction.

---

4 As of December 2006, Japan claimed that seventeen individuals had been abducted by the DPRK in the 1970s and the 1980s, including those thirteen individuals that the DPRK admitted having abducted.

5 Narushige Michishita, "Kitachosen no kakumondai," in *Chosenhanto wo meguru kongo no kokusaikankei no tenbo,* ed., Kokusai Kinyu Joho Senta (Tokyo: Kokusai Kinyu Joho Senta, 2005), p. 55; *Asahi Shimbun*, July 13, 2006.

6 There is a view that DPRK already possesses some operational nuclear warheads. Keisuke Ebata, *Nihon boei no arikata* (Tokyo: KK Best Sellers, 2004), pp. 164-173; David S. Cloud and David E. Sanger, "North Korean threat worsens: Nation now seen as able to put nuclear warheads on missiles," *San Francisco Chronicle*, April 29, 2005, p. A-1.

Japan's Response to the Second Nuclear Crisis

This section examines Japan's response to the nuclear crisis that started after the U.S. revelation in October 2005 of the DPRK's engagement in uranium enrichment.

Japan's initial response to the surfacing of the DPRK's uranium enrichment program indicated its desire to prevent escalation of the new nuclear crisis. Soon after disclosing Pyongyang's uranium enrichment program in October 2002, the U.S. decided to repeal the 1994 Agreed Framework and stopped the provision of heavy fuel oil to the DPRK as was stipulated under the Agreed Framework. The U.S. wanted to stop the provision immediately, but Japan and the Republic of Korea (ROK or South Korea) were opposed.[7] Consequently, the portion of heavy fuel oil allocated for November 2002 was provided. Yet the ROK and Japan could not change Washington's strong determination to freeze the oil provision and agreed on its freeze at a board meeting of the Korean Peninsula Energy Development Organization (KEDO) on November 14, 2002.[8] The termination of the U.S. heavy oil supply led the DPRK to resume its operation of nuclear facilities. Japan and the ROK anticipated that termination of heavy oil supply to the DPRK would lead to the collapse of the Agreed Framework and the resumption of the DPRK's activity of reprocessing spent fuel rods to extract plutonium.[9]

Yet, this does not mean that Japan completely opposed the U.S. hardline policy toward the DPRK. In fact, Tokyo requested Washington to maintain the option of a military action against Pyongyang.[10] Also, during the Japan-U.S.-ROK trilateral talks held on August 13-14, 2003, Japan asked the U.S. not to promise the DPRK that it would not attack North Korea with nuclear weapons.[11]

Before the start of the six-party talks on August 20, 2003, Japan's Ministry of Foreign Affairs (MOFA) decided its basic policy for the talks, which subsequently received approval of the Prime Minister's Office.[12] According to the basic policy, Japan planned to make the following demands at the talks: (1) the DPRK should immediately stop its reprocessing of spent fuel rods, abandon existing nuclear weapons, dismantle its nuclear facilities, and accept nuclear inspections. The DPRK should abandon its nuclear programs verifiably and irreversibly; (2) the DPRK should stop its development, experiments, and export of ballistic missiles, and remove and discard its deployed ballistic missiles; and (3) the DPRK should fully resolve the issue of the abduction of Japanese nationals.

---

7 *Asahi Shimbun*, November 14, 2002, evening edition; *Nihon Keizai Shimbun*, November 14, 2002.

8 KEDO was established in 1995 to provide two light-water reactors (LWRs) to the DPRK.

9 Not only leftist scholars but also mainstream scholars in Japan hold the view that the Bush administration's hardline poicy is partly responsible for the collapse of the Agreed Framework and North Korea's nuclear test in October 2006. Masao Okonogi, "Kitachosen no kakujikken wo yurushitamono," *Ronza* (December 2006), p. 27; Akihiko Tanaka, "Kitaidekinai Kitachosen no kaku haizetsu," *Ronza* (December 2006), p. 40.

10 *Tokyo Shimbun*, July 31, 2003.

11 *Yomiuri Shimbun*, August 22, 2003.

12 *Yomiuri Shimbun*, August 24, 2003.

The basic policy also specified that the following actions be taken after progress on the nuclear and abduction issues: (1) Japan will consider the provision of heavy oil and then electricity to the DPRK, possibly in the form of building thermal power plants or transmission of electricity from the ROK, if KEDO terminates its project of building two LWRs in the DPRK. It will also consider the provision of humanitarian assistance, mainly in the form of food aid; (2) Japan will provide economic assistance to the DPRK after diplomatic normalization; and (3) Japan will consider the issue of guaranteeing the survival of the current regime of the DPRK. According to the basic policy, Japanese negotiators would not mention potential assistance to North Korea. As stated in the Pyongyang Declaration issued at the first Japan-North Korea summit of September 2002, Japan would only state that it would provide economic assistance only after the success of diplomatic normalization, unless the DPRK makes positive responses on the abduction issue and other outstanding issues.

The First Round of the Six-Party Talks

On August 27-29, 2003, the first round of the six-party talks convened in Beijing. At the talks, Japan mostly followed the aforementioned basic policy. Japan demanded that North Korea should abandon all nuclear weapons development programs completely, irreversibly, and verifiably.[13] With regard to economic incentives for North Korea, Japan took the stance that it would discuss energy assistance to the country in detail at an appropriate time if North Korea took concrete steps toward nuclear abolishment.[14] However, Japan told the DPRK that economic assistance would be provided only after bilateral diplomatic normalization and that the resolution of not only the nuclear issue but also missile and abduction issues must precede diplomatic normalization.[15] The U.S. demanded complete, verifiable, and irreversible dismantlement (CVID) of the DPRK nuclear programs and offered no economic incentives.

The ROK took a more compromising stance, proposing a three-stage resolution.[16] The first stage consisted of the DPRK's declaration to abandon its nuclear programs and the declaration by the other five countries to guarantee the DPRK's security. The second stage consisted of the DPRK's freezing of its nuclear programs, its acceptance of inspections, and energy assistance by the ROK, the People's Republic of China (PRC or China), and Russia. The third stage consisted of the complete dismantling of the DPRK's nuclear programs, documentation of the security guarantee, and diplomatic improvement between the DPRK and the U.S. The ROK's proposal indicated its very limited expectations for economic concessions to the DPRK by the U.S. and Japan.

13 Gaimusho, *Rokusha-kaigo (gaiyo to hyoka)*, n.d., online at <http://www.mofa.go.jp/mofaj/area/n_korea/6kaigo/6kaigo_gh.html>.

14 Ibid.

15 Ibid.

16 *Asahi Shimbun*, February 20, 2004, evening edition.

North Korea, for its part, proposed a four-stage, give-and-take process.[17] The four-stage process consisted of the following components: (1) The U.S. would resume provision of heavy fuel oil to North Korea and would extend humanitarian aid to North Korea, while North Korea would declare the abandonment of its nuclear weapons and programs; (2) Once the U.S. concludes a non-aggression treaty with North Korea and compensated North Korea's energy loss, North Korea would freeze its nuclear facilities and nuclear materials and accept inspections; (3) At the same time, the U.S. and Japan would normalize their diplomatic relations with North Korea, and North Korea would conclude an agreement on its missiles with the two countries; and (4) Upon the completion of the construction of LWRs, North Korea would dismantle its nuclear facilities.

The first round of the talks clearly showed wide gaps between North Korea and the U.S. and Japan. The first round ended with no tangible results.

After the first round of the six-party talks, Japan took inconsistent actions toward North Korea. At the third meeting of the Proliferation Security Initiative (PSI) on September 3, 2003 held in Paris,[18] Japan and European countries opposed a U.S. proposal to adopt a PSI declaration to designate North Korea as a country of proliferation concern.[19] Japan and other participants also disagreed with another U.S. proposal to enact a new law enabling inspections of ships belonging to non-member countries on the high seas.[20] Japan and the U.S. disagreed over the participation of the PRC and the ROK in the PSI. While Japan was in favor of their participation, the U.S. opposed it on the grounds that it would make the PSI organization as immobile as the UN.[21]

While it indicated its reluctance to heighten tensions with North Korea at the PSI meeting, Japan enacted a new law to pave the way for future economic sanctions against North Korea. On February 9, 2004, the Japanese Diet passed legislation to revise the Foreign Exchange and Foreign Trade Law, which became effective on February 26. It enabled the government to unilaterally impose economic sanctions on a country by freezing remittance from Japan to the country and Japan's trade with the country, if deemed necessary for maintaining the peace and security of Japan. Previously, imposition of economic sanctions required a UN resolution or a bilateral agreement. The primary objective of the revision was to give the Japanese government a more effective diplomatic card against the DPRK, particularly to promote the resolution of the abduction issue.[22] A freeze on remittance to Pyongyang could have a major impact on the country because the annual remittance was estimated at as much as several hundred million dollars,[23] although the legal remittance amounted

17 *Tokyo Shimbun*, August 29, 2003, evening edition.

18 Washington initiated the PSI in May 2003 as a more proactive and effective measure to prevent proliferation transfers of the WMDs and related technologies to or from nation states and non-state actors of proliferation concern.

19 *Asahi Shimbun*, September 11, 2003.

20 Ibid.

21 Ibid.

22 *Sankei Shimbun*, January 29, 2004.

23 *Yomiuri Shimbun*, November 21, 2003. A former member of the Chosen Soren, the pro-DPRK General Association of Korean Residents in Japan, gave an estimate of about

to about 2.7 billion yen in 2003 according to a survey conducted by the Ministry of Finance in 2005.[24]

The Second Round of the Six-Party Talks

The second round of the six-party talks was held on February 25-28, 2004 in Beijing. Japan hardly diverged from its previous policy stance. Japan, along with the U.S. and the ROK, demanded that the DPRK abandon all of its nuclear programs and admit to the existence of its uranium enrichment program.[25] Furthermore, in its keynote address, Japan stressed the importance of resolving the abduction issue, which it saw as the biggest issue between Japan and the DPRK.[26] On its part, the DPRK denied its engagement in uranium enrichment and maintained that its peaceful uses of nuclear technology should be permitted.[27] On February 26, the DPRK maintained that no breakthrough toward the resolution of the nuclear problem could be achieved because the U.S. demanded that the DPRK abandon its nuclear programs first, including peaceful uses of nuclear energy and that the DPRK would have made additional offers if the U.S. had offered concrete measures in return for the DPRK's freeze on nuclear activities.[28] Japan did not attempt to narrow the gap between the DPRK and the U.S. and maintained a stance close to that of the U.S. The second round of the talks ended with little progress.

Revision to the Japan-U.S. ACSA While the second round of the six party talks was under way, Japan and the U.S. made a new security arrangement that enhanced their readiness for military contingencies on the Korean peninsula. On February 27, 2004, they signed a revised Acquisition and Cross-Servicing Agreement (ACSA) that enabled Japan's Self Defense Forces (SDF) to provide ammunitions to U.S. forces coping with the so-called "Shuhenjitai," situations in areas surrounding Japan that seriously affected Japan's national security, such as military contingencies on the Korean peninsula. After the second round of the six-party talks, Japan did not make any significant efforts to resolve the nuclear issue and continued to give a higher priority to the resolution of the abduction issue. Furthermore, it adopted new measures to improve its readiness to deal with military contingencies in areas surrounding Japan, which can involve the DPRK.

After the second round of the Six-Party talks, Japan did not make any significant efforts to resolve the nuclear issue and continued to give a higher priority to the resolution of the abduction issue. Furthermore, it adopted new measures to improve

---

$100 million per year at most. Lee Chek, *Gekishin! Chosen soren no uchimaku* (Tokyo: Shogakukan, 2003), pp. 58-66.

24 *Kyodo News*, February 12, 2005, online edition.

25 Gaimusho, *Dainikai rokusha-kyogi (gaiyo to hyoka)*, March 1, 2004, online at <http://www.mofa.go.jp/mofaj/area/n_korea/6kaigo/6kaigo2_gh.html>.

26 Ibid.

27 Ibid.

28 *Asahi Shimbun*, February 27, 2004; *Yomiuri Shimbun*, February 27, 2004.

its readiness to deal with miltary contingencies in areas surrounding Japan, including the Korean peninsula.

The Second Japan-DPRK Summit The second Japan-DPRK summit between Prime Minister Junichiro Koizumi and Kim Jong-Il was held on May 22, 2004. Koizumi's primary objective was to bring to Japan the family members of the five abductees who had returned to Japan in October 2002. He succeeded in achieving this objective when two abducted Japanese couples returned with their children to Japan in October 2002 following the first summit in September 2002. He also succeeded in getting Kim's promise to continue the DPRK's moratorium on the test-launching of ballistic missiles. In return for Kim's concessions, Koizumi promised to provide food aid worth 250,000 tons and medical aid worth $10 million. On the nuclear issue, Koizumi reiterated Japan's position that complete denuclearization and subsequent international verification would be necessary. According to him, Kim stated that his objective was the denuclearization of the Korean peninsula and expressed his desire to peacefully resolve the nuclear issue through the six-party talks. Thus, the second summit also resulted in little substantive progress on the nuclear issue.

The law restricting the entry of specific foreign ships Although the second summit somewhat improved its relations with the DPRK, Japan made a new action that soured those relations once again. On June 14, 2004, the Diet passed legislation that would enable the Japanese government to ban the entry of ships belonging to a specific country and those ships that visited the country if deemed necessary for maintaining the peace and security of Japan. This new law was intended primarily to strengthen Japan's bargaining power vis-à-vis North Korea.[29] The enforcement of this law had the potential of inflicting significant economic damages on the DPRK because its trade with Japan depended mostly on ships for transportation of goods and also because most remittance from Japan had been shipped.[30] Additionally, Japan changed defense-related laws, which further aggravated its relations with North Korea.

The law for smooth operations of the U.S. forces On June 14, 2004, the Diet enacted a law to facilitate smooth military operations of U.S. forces dealing with actual and expected armed attacks on Japan. The law came into effect at the same time as the revised ACSA to achieve its purpose. Further, the law accompanied an amendment to the SDF Law that enabled the SDF to provide goods and services to U.S. forces at the time of both actual and expected armed attacks on Japan, even before the SDF received an order for mobilization to conduct defense operations.

The Military Material Maritime Transport Restriction Law On June 14, 2004, the Diet enacted another defense-related law that allowed the Maritime SDF (MSDF) to restrict maritime transport of military materials within the Japanese territorial waters and the high seas during military attacks on Japan. Apparently, Japan made these legal changes involving military contingencies with the DPRK in mind.

29 *Sankei Shimbun*, April 6, 2004.

30 *Kyodo News*, February 12, 2005, online edition.

The Third Round of the Six-Party Talks

The third round of the six-party talks began on June 23, 2004 in Beijing and lasted for four days. This round focused on the first stage of nuclear dismantlement, that is, the freezing of the DPRK's nuclear program. This time, Japan softened its stance toward the DPRK, due probably to positive results on the abduction, missile, and nuclear issues at the second Japan-DPRK summit of May 2004.

Japan expressed its willingness to take part in energy assistance to North Korea in the form of financial contributions on the conditions that all of North Korea's nuclear programs, including uranium enrichment, should be frozen and the DPRK should disclose and verify relevant information on the freeze.[31] However, Japan restated its previous position that it would provide economic assistance to the DPRK only after the two countries achieved diplomatic normalization, which was conditional on the comprehensive resolution of the nuclear, missile, and abduction issues between them.[32]

While Japan somewhat softened its stance on the provision of energy assistance to North Korea, the U.S. did not change its stance. On June 23, the U.S. proposed that Japan, the ROK, the PRC, and Russia provide the DPRK with heavy fuel oil and the U.S. offer a "provisional" guarantee of the DPRK security if the DPRK provides a full declaration and cessation of its nuclear activities, secures any fissile materials, disables any dangerous materials, and allows inspectors to return.[33] Thus, the U.S. maintained its previous stance of withholding economic concessions to the DPRK. For its part, North Korea demanded the provision of enough heavy oil to fuel thermal energy plants to produce 2 megawatts of electricity per year, U.S. participation in energy assistance, the lifting of U.S. economic sanctions against the DPRK, and the declassification of the DPRK as a state sponsoring terrorism.[34]

The U.S., along with Japan and the ROK, was mainly interested in reaching an agreement to have the DPRK abandon its nuclear programs, while the DPRK gave priority to compensation measures for the freeze of its nuclear programs.[35] Understandably, the DPRK and other participants, particularly the U.S., could not agree on the measures for the first stage of denuclearization.[36] Thus, this round of the six-party talks produced few meaningful results, although Japan somewhat eased its stance on the provision of energy assistance to the DPRK.

After the third round of the six-party talks, Japan's relations with North Korea briefly improved over the abduction issue, but deteriorated again due to the North Korea's poor handling of the issue and changes in Japan's security policy.

---

31 Gaimusho, *Kitachosen no kakumondai nikansuru dai3kai rokusha-kaigo (gaiyo to hyoka)*, June 27, 2004, online at <http://www.mofa.go.jp/mofaj/area/n_korea/6kaigo/6kaigo3_gh.html>; *Asahi Shimbun*,>. June 25, 2004.

32 Ibid.

33 *CNN Online*, June 23, 2004.

34 *Asahi Shimbun*, June 25, 2004.

35 Gaimusho, *Kitachosen no kakumondai nikansuru dai3kai rokusha-kaigo (gaiyo to hyoka)*.

36 Ibid.

In July 2004, there was another progress on the abduction issue as a result of the second Japan-DPRK summit. The DPRK allowed the American husband, Charles Jenkins, and two children of Hitomi Soga, who was abducted but returned to Japan in October 2002 following the first summit, to go to Japan. Having first reunited in Indonesia on July 9, they arrived together in Japan on July 18. The DPRK also reported the results of its investigations on the allegedly dead Japanese abductees, which it had promised to conduct at the second Japan-DPRK summit. In November 2004, the DPRK returned the remains of two abducted Japanese citizens, who allegedly died in the DPRK. It also submitted to Japan the death certificates of the eight abducted Japanese citizens. Yet, Japan found that these remains were not those of the two Japanese abductees and that the certificates were unreliable. As a result, the tension between Japan and the DPRK reached a new height.

Meanwhile, changes in Japan's security policy stance aggravated its relations with North Korea. Between October 25-27, 2004, Japan hosted for the first time a PSI maritime interdiction training exercise, which was aimed at the interdiction of the shipment of weapons of mass destruction (WMDs) by problematic states such as the DPRK. On December 20, 2004, Japan adopted a revised National Defense Program Outline (NDPO), which stressed the security threat posed by the DPRK and clearly stated Japan's decision to deploy, develop, and operate a ballistic missile defense (BMD) system in cooperation with the U.S., apparently as a countermeasure to North Korea's ballistic missiles.

On February 10, 2005, the DPRK declared the possession of nuclear weapons and thereby heightened the level of the existing international tension and aggravated its relations with Japan. The tension between Japan and the DPRK escalated further as a result of Japan's recognition of Minoru Tanaka as the 16th Japanese citizen abducted by the DPRK on April 27, 2005 and its demand for his return.

The Fourth Round of the Six-Party Talks

About a year after the third round, the fourth round of the six-party talks convened from July 26-August 7, 2005. The second session was held from September 13-19. In this round, Japan did not make any significant new proposals. In its keynote speech, Japan pointed to the threat of North Korea's ballistic missiles to the peace and stability of the Northeast Asian region and the world as a whole.[37] It also urged the DPRK to agree to the complete abandonment of all its nuclear programs, including uranium enrichment, under reliable international verification.[38] Japan expressed its readiness to commit itself to the provision of a "security guarantee" to the DPRK in return for the DPRK's commitment to abandoning its nuclear programs.[39] Yet it did not change the policy to demand a comprehensive resolution of the nuclear, missile, and abduction issues as a precondition for diplomatic normalization with the

37 Gaimusho, *Dai4kai rokushakaigo: Sasae taishi niyoru kichohatsugen youshi*, July 27, 2005, online at <http://www.mofa.go.jp/mofaj/area/n_korea/6kaigo/6kaigo4_kh.html>.

38 Ibid.

39 Ibid.

DPRK.[40] With regard to energy assistance, Japan welcomed the ROK's proposal to provide electricity worth 2 megawatts directly from the ROK.[41]

For its part, the DPRK slightly changed its previous stance. It insisted on the right as a sovereign state to peaceful uses of nuclear energy and demanded the resumption of the construction of LWRs.[42] Also, as before, it squarely denied the existence of a uranium enrichment program.[43] On July 22, the (North) Korean Central News Agency (KCNA) quoted the spokesman of the DPRK's foreign ministry as saying that it is necessary to terminate the Korean War and that conclusion of a peace treaty between the DPRK and the U.S. is indispensable to the resolution of the nuclear issue.[44] On July 27, North Korea clearly rejected the U.S. proposal made at the third round of the six-party talks and stated that it would publicly promise to abandon its nuclear weapons and programs in a verifiable manner if the nuclear threat by the U.S. was removed and if the relations between North Korea and the U.S. were normalized.[45] On the same day, the DPRK also demanded removal of the U.S. nuclear umbrella (apparently for the ROK) and a ban on bringing nuclear weapons into the Korean peninsula.[46] On July 31, the DPRK demanded the resumption of the KEDO project of constructing LWRs, which had been halted in late 2003, on the grounds that electricity transmission from the ROK could be stopped unilaterally by the ROK.[47] For its part, on July 29 the U.S. reiterated its policy of opposing the North's peaceful uses of nuclear energy.[48]

Although there were still many differences of opinion, the six parties succeeded in producing their first joint statement, in which they agreed to take coordinated steps to implement the process of the DPRK's denuclearization in a phased manner in line with the principle of "commitment for commitment, action for action." Judging from the joint statement, one can see some softening of the U.S. stance toward the DPRK. The U.S. adopted a more positive stance toward energy assistance to the North and gradual, phased implementation of denuclearization that would involve the occasional provision of economic incentives to the DPRK. This softening of the U.S. stance toward Pyongyang might have been affected by the earlier softening of Japan's stance. However, the extent of Japan's impact is unclear and seems smaller than that of the ROK's.

With regard to the contentious issue of the provision of LWRs, the joint statement used very vague wordings. Needless to say, the U.S. opposed the idea. Japan, in its closing speech, clarified its interpretation of "an appropriate time" for discussing the subject of the provision of LWRs to the DPRK. Japan insisted that

---

40 Ibid.

41 Ibid.

42 Gaimusho, *Dai4kai rokushakaigo (gaiyo to tenbo)*, September 2005, online at <http://www.mofa.go.jp/mofaj/area/n_korea/6kaigo/6kaigo4_gh.html>.

43 Ibid.

44 *Yomiuri Shimbum*, July 26, 2005.

45 *Mainichi Shimbun*, July 28, 2005.

46 Ibid.

47 *Tokyo Shimbun*, July 31, 2005.

48 Remarks by a spokesman of the Department of State. *Tokyo Shimbun*, July 30, 2005, evening edition.

the discussion would start only after the DPRK abandoned all of its nuclear weapons and existing nuclear programs under credible international inspections, completely abided by all the international agreements and norms regarding the uses of nuclear energy that include the NPT and IAEA safeguards, and thereby earned the trust of the international society.[49] Thus, Japan indicated its willingness to recognize the DPRK's right to peaceful uses of nuclear technology, which was in contrast to the U.S. position. Yet, Japan took the same stance as the U.S. that LWRs would not be provided in return for the DPRK's denuclearization.

After the fourth round of the six-party talks, the tension between the U.S. and the DPRK reached a new height because of U.S. financial sanctions imposed in September 2005 on a Macao bank, Banco Delta Asia (BDA), which was accused of being involved in money laundering for the DPRK. Japan by no means opposed the U.S. action, but during his talks with Secretary of State Condoleezza Rice on October 28, 2005, Foreign Minister Nobutaka Machimura urged her to discuss economic assistance to the DPRK in the next round of the six-party talks by saying, "DPRK's participation in the next round of the talks will be unlikely if economic assistance is not to be discussed."[50] This indicated Japan's increased willingness to give economic concessions to the DPRK for its denuclearization.

The First Session of the Fifth Round of the Six-Party Talks

From November 9-11, 2005, the first session of the fifth round of the six-party talks convened. The session ended with little progress. In this round, Japan proposed the creation of two working groups to discuss the issue of abandoning and verifying nuclear programs and the issue of economic and energy assistance.[51] But Japan did not make new substantive proposals of any significance. It restated the importance of resolving the abduction issue, and then stressed that improving Japan-DPRK relations by resolving outstanding issues such as the abduction issue would have a positive impact on the progress of the six-party talks and that progress on nuclear and missile issues at the six-party talks would improve the prospects for diplomatic normalization between Japan and the DPRK.[52]

The DPRK strongly criticized recent U.S. policies, especially financial sanctions on it imposed in September 2005.[53] DPRK argued that such an action violated the joint statement adopted at the fourth round of the six-party talks, that the action made it impossible for the DPRK to follow through on its promises made in the statement, and that the U.S. should revise its hostile policy toward the DPRK.[54]

---

49 Gaimusho, *Dai4kai rokushakaigo heimakushiki (Sasae dancho hatsugen)*, September 2005, online at <http://www.mofa.go.jp/mofaj/area/n_korea/6kaigo/6kaigo4_hei.html>.

50 Sekai henshubu, "Kitachosen no 5 dankai teian," *Sekai* (January 2006), p. 198.

51 Ibid., p. 200.

52 Japan made these points to the DPRK during its talks with the DPRK at a dinner party on November 8 and a bilateral talk on November 10. Gaimusho, *Dai5kai rokushakaigo dai1ji kaigo (gaiyo to tenbo)*, November 2005, online at <http://www.mofa.go.jp/mofaj/area/n_korea/6kaigo/6kaigo5_gt.html>.

53 Ibid.

54 Ibid.

While this session aimed at compiling concrete plans, measures, and procedures for implementing the agreement in the joint statement,[55] the strong animosity between the DPRK and the U.S. resulting from the U.S. financial sanctions prevented meaningful discussion on the implementation of the joint statement, including the creation of working groups proposed by Japan.

After the first session of the fifth round of the talks, the tension between the U.S. and the DPRK mounted further as a result of KEDO's decision in November 2005 to terminate its project to construct two LWRs in the DPRK, which had been suspended since the end of 2003. The decision virtually meant the end of KEDO. Japan's relations with the DPRK also worsened, due partly to the failure of their bilateral talks and Japan's adoption of more hardline measures against the DPRK.

After the first session of the fifth round of the talks, Japan pursued the policy of promoting the progress of the six-party talks and the Japan-DPRK bilateral talks by "appropriately linking the two."[56] Following this policy, Japan proposed that Japan and the DPRK hold comprehensive talks to discuss the abduction, nuclear, missile, and normalization issues, and succeeded in gaining Pyongyang's agreement. Then, the first round of the comprehensive talks was held in Beijing from February 4-8, 2006. They discussed the abduction issue on February 4, the normalization issue on February 6, and the nuclear and missile issues in the morning of February 7. This order suggested Japan's priorities. In fact, Japan publicly stated that the abduction issue was the issue of the highest priority.[57] Throughout the talks, Japan repeated its basic position that it would not normalize its diplomatic relations with the DPRK without the resolution of the outstanding issues between the two countries.[58] Japan did not make any substantial, new proposals for the resolution of the nuclear issue. It urged the DPRK to come back to the negotiating table of the six-party talks without any conditions. The DPRK responded by restating its position that its return to the talks was conditional on the lifting of recent U.S. financial sanctions on it.[59] No noticeable progress was made on any of the outstanding issues during the bilateral talks.[60] Worse, bilateral relations deteriorated further as a result of Japan's suggestion of imposing economic sanctions on the DPRK. To this, the DPRK indicated that it would not care about Japan's sanctions because of its strong trade ties with the PRC and the significant decline in its trade with Japan that had already taken place.[61]

It also told Japan that it would resort to "potent physical countermeasures" to any sanctions by Japan.[62] Having gained few concessions on the abduction issue, on February 16, Japan adopted a new policy of increasing economic pressure on the DPRK by more strictly applying existing laws to economic transactions with the DPRK and to the *Chosen Soren* (the pro-DPRK General Association of Korean

55 Ibid.

56 Ibid.

57 Gaimusho, *Dai1kai niccho hokatsu heiko kyogi (gaiyo)*, February 2006, online at <http://www.mofa.go.jp/mofaj/area/n_korea/abd/hokatsu.html>.

58 Ibid.

59 Ibid.

60 Ibid.

61 *Kyodo News*, February 23, 2006, online edition.

62 Ibid.

Residents in Japan).[63] It also began to consider imposing financial sanctions on the DPRK-related financial institutions, as Washington did to BDA.[64]

Additionally, the ruling Liberal Democratic Party (LDP) stepped up its effort at enacting a North Korea Human Rights Law, a Japanese version of U.S.'s North Korea Human Rights Act of 2004, which would enable the government to impose economic sanctions against the DPRK based on the revised Foreign Exchange and Foreign Trade Law of 2004 and the Law Restricting the Entry of Specific Foreign Ships of 2004, if progress on the abduction issue was deemed insufficient. On February 16, 2006, the committee of the LDP in charge of the abduction issue decided the basic outline of the new law.[65] While preparation for the law was under way, on February 23, 2006, the government issued warrants of arrest for two DPRK agents accused of abducting Japanese couples and demanded that the DPRK hand them over. Japan enacted the law on June 12, 2006, which then came into effect on June 23. Besides putting more pressure on the DPRK on the abduction issue, the law aimed at improving public awareness and understanding of the abduction issue and other human rights violations by the DPRK.[66] Not surprisingly, the DPRK strongly criticized the new law.[67] The tension between Japan and the DPRK further escalated as a result of the DPRK's missile tests and nuclear test.

Response to North Korea's Missile Tests

On July 5, the DPRK test-fired seven missiles. One of the missiles was believed to be a Taepodong-2 missile, while the rest were Scud missiles and Nodong missiles. Considering the DPRK's ballistic missiles as a major threat, Japan made quick and resolute reactions to the tests. It immediately imposed unilateral sanctions on the country and actively promoted the adoption of a sanction resolution by the UN Security Council.

On July 5, immediately after the missile tests, Japan announced nine countermeasures, which included: banning of (1) the entry of the Man Gyong Bong 92, North Korea's primary cargo and passenger vessel to Japan; (2) the landing of crews of other North Korean ships; (3) the entry of charter flights from the DPRK; (4) the entry of the DPRK officials; and (5) the re-entry of Korean residents to Japan who work for the DPRK government. The DPRK strongly criticized Japan's sanctions and stated that its relations with Japan had gone beyond the worst point, reaching a stage of confrontation, and became even worse than its relations with the U.S.[68]

In the afternoon of July 5, Japan presented to an unofficial ambassador-level meeting of the UN Security Council a draft of a binding resolution based on Chapter

63 *Sankei Shimbun*, February 17, 2006.

64 Ibid.

65 *Yomiuri Shimbun*, February 17, 2006.

66 The law designated a week from December 19 to 16 as "the North Korea Human Rights Violation Problem Awareness Week."

67 *Chosen Shimpo*, June 29, 2006.

68 Remarks of DPRK ambassador Song Ilho on July 7. *Sankei Shimbun*, July 7, 2006; *Chosun Ilbo*, July 8, 2007, online Japanese edition.

VII of the UN Charter that would require member states to impose economic sanctions on the DPRK in the form of banning their supply of funds, materials, and technologies that could help North Korea's missile and other WMD programs.[69] Considering that missile tests are not prohibited by international laws, it was not a surprise that Resolution 1695 that the UN Security Council adopted on July 15 did not include sanctions based on Chapter VII. Yet, Japan's draft exemplified its strong reaction.

On September 19, Japan imposed additional sanctions on the DPRK, designating fifteen institutions and one individual as related to the DPRK's missile and other WMD programs and restricting their financial transactions on the basis of the revised Foreign Exchange and Foreign Trade Law of 2004. On September 23, North Korea's Central Broadcast severely criticized the financial sanctions as a provocative act that followed the hostile U.S. policy toward the DPRK.[70]

Japan also made a military response to the DPRK's missile tests. On July 6, the Japan Defense Agency (JDA) decided to accelerate the construction of a BMD system that was originally scheduled for completion by 2011. The decision was later endorsed by the government.[71] On July 7, the government decided to complete the deployment of four units of PAC-3 missiles in 2007, which was originally scheduled for the end of March 2008.[72] In August, JDA decided to request a budget of 219 billion yen for missile defense in the Fiscal Year (FY) 2007, which is approximately a 50 percent increase from FY 2006.[73] While accelerating the construction of a BMD system, Japan attempted to improve its satellite surveillance capability. On September 11, it successfully launched its third "Information Gathering Satellite (IGS)."[74] It plans to launch a fourth IGS in early 2007, which would grant Japan the capacity to monitor any spot on earth once a day. On September 16, 2006, the LDP revealed a final draft of a space law that would nullify the Diet Resolution of 1969 that permitted the use of space only for peaceful purposes and thereby would allow Japan to possess advanced military satellites.[75]

---

69 *Nihon Keizai Shimbun*, July 6, 2006, evening edition.

70 *Nihon Keizai Shimbun*, September 23, 2006.

71 *Jiji Tsushin*, July 6, 2006, online edition.

72 The first unit of PAC-3 missiles is scheduled for deployment by the end of March 2007. *Yomiuri Shimbun*, July 8, 2006.

73 *Asahi Shimbun*, August 23, 2006. The draft budget compiled by the Ministry of Finance in December allocated 182.6 billion yen for the missiles defense, which was about a 30 percent increase from FY 2006. *Asahi Shimbun*, December 20, 2006. Japan's FY starts in April.

74 IGSs are de facto spy satellites and managed by the Cabinet Office, not by JDA. Their resolution cannot go beyond that of civilian satellites. *Sankei Shimbun*, December 31, 2005. Japan has spent 505 billion yen for the IGSs since 1998. *Tokyo Shimbun*, September 12, 2006.

75 *Yomiuri Shimbun*, September 17, 2006. The review of the resolution began earlier. In December 2005, Fukushiro Nukaga, head of JDA, former heads of JDA, defense industry representatives, scholars, formed an unofficial committee to review it. In January 2006, the LDP decided to complete its review of the resolution and make a proposal to the government by the summer. In June 2006, LDP's special sub-committee on space development compiled

Voices within the government for the acquisition of the military capability to strike military bases in a hostile state became stronger.[76] On July 9, the head of JDA Fukushiro Nukaga said that it would be natural for Japan, as a sovereign state, to have a limited offensive capability in order to protect Japanese citizens, while on the same day Foreign Minister Taro Aso maintained that Japan could not remain idle if nuclear missiles were aimed at it.[77] Also, then-Chief Cabinet Secretary Shinzo Abe stressed the need to seriously discuss the issue of acquiring the capability to strike military bases in a hostile state.[78] Japan's series of responses to the DPRK's missile tests heightened, rather than eased, the tension between the two countries. Unfortunately, the bilateral relations deteriorated further as a result of the DPRK's nuclear test.

Against this backdrop of a mounting tension between Japan and North Korea, Shinzo Abe became prime minister on September 26, 2006, succeeding Junichiro Koizumi. He was well-known for his tough stance toward the DPRK. He created the new position of a minister in charge of the abduction issue and appointed his Chief Cabinet Secretary Yasuhisa Shiozaki. He also established a new section within the Cabinet in charge of the abduction issue on September 29, 2006, which consisted of cabinet members and headed by the prime minister. A renewed emphasis on the abduction issue by the Abe administration cast a shadow over the already strained Japan-DPRK relations.

Response to North Korea's Nuclear Test

On October 9, 2006, the DPRK announced a successful underground nuclear test, which was carried out despite strong international calls for the DPRK's restraint after the North's announcement on October 3 of the intention to conduct a nuclear test. Japan reacted strongly to the test, severely criticizing the action, quickly imposing unilateral sanctions on the DPRK, and actively pursuing the adoption of a sanction resolution by the UN Security Council.

On October 9, after the U.S. presented a draft resolution to a UN Security Council meeting, Japan submitted additional proposals. The U.S. draft called for economic sanctions based on Chapter VII of the UN Charter that included a ban on the provision of funds, goods, and technology that could contribute to the DPRK's WMD-related programs. Japan's proposals required member states to prohibit (1) the entry of all North Korean ships and landing and taking-off of all the North Korean flights, (2) all imports from North Korea, and (3) the entry of North Korean high-ranking officials.[79] Apparently, the objective of its proposals was to supplement

---

an outline of a space law. *Sankei Shimbun*, December 31, 2006; *Yomiuri Shimbun*, January 6, 2006; *Sanyo Shimbun*, June 2 2006.

76 For support for such an action outside the government, see for instance: Sekai Heiwa Kenkyusho, *21 seiki no kokkazo nitsuite* (Tokyo: Sekai Heiwa Kenkyusho, 2006), p. 5; Jun Sakurada, "'Nihon kakubuso' wa anzenhoshorongi no adabana dearu," *Chuokoron* (June 2003), pp. 106-15.

77 *Shimbun Akahata*, July 11, 2006.

78 Ibid.

79 *Asahi Shimbun*, October 10, 2006.

the sanctions proposed by the U.S. The final resolution, Resolution 1718, adopted by the UN Security Council on October 14, mostly reflected the U.S. draft, but it did not incorporate the aforementioned Japanese proposals, which were apparently deemed too harsh. While the resolution debate was under way, on October 11 Japan announced that it would unilaterally implement basically the same measures against the DPRK that it proposed at the UN Security Council meeting, banning the entry of all North Korean ships for six months, all imports from the DPRK for six months, and the entry of North Koreans for six months.

Also, in response to the U.S. draft that called for inspections of cargo to or from the DPRK, Japan showed its willingness to contribute to ship inspections. Japan can conduct ship inspections only when it faces "Shuhenjitai"; extraordinary situations in areas surrounding Japan that seriously affect Japan's peace and security, such as military contingencies on the Korean Peninsula. In order to provide legal grounds for its ship inspections and its support of U.S. ship inspections, on October 11 the Abe administration began a discussion of designating the situation resulting from the DPRK's nuclear test as a "Shuhenjitai."[80] On October 15, Foreign Minister Taro Aso clearly stated that Japan should make such a designation.[81] On October 12, the ruling LDP also started a discussion about the enactment of a new special law that would enable Japan to support ship inspections by countries other than the U.S.[82] At the same time, Japan decided to accelerate the construction of a BMD system. On October 12, it decided to reschedule the first deployment of SM-3 missiles by 3 months from by the end of March 2008 to by the end of 2007.[83]

Furthermore, after the nuclear test, the discussion of the possibility of going nuclear became lively in Japan. It was strongly promoted by the remarks of Shoichi Nakagawa, chairman of the Policy Research Council of the LDP,[84] on October 15 that a debate over whether to go nuclear was necessary, that the constitution does not prohibit the possession of nuclear arms, and that the possession of nuclear weapons may reduce the risk of being attacked.[85] On October 16, Prime Minister Abe, who is also the president of the LDP, publicly denied the intention of the government to support the idea of going nuclear or even officially discuss it, but stopped short of dismissing Nakagawa.[86] Despite Abe's remark, on October 18, Foreign Minister Aso expressed in the Diet a support for the idea of discussing various issues including the option of going nuclear.[87] During this time, soon after North Korea's announcement of its intention to conduct a nuclear test, on October 7, Sankei Shimbun, one of

---

80 *Sankei Shimbun*, October 12, 2006. For more on this issue, see Tetsuo Maeda, "Shuhenjitai to senpaku rinken," *Sekai* (December 2006), pp. 73-77.

81 *Nihon Keizai Shimbun*, October 15, 2006. As of this writing, the final decision has not been made, due partly to reservations expressed by JDA and ruling party members to such a designation. *Mainichi Shimbun*, October 14, 2006.

82 *Chunichi Shimbun*, October 13, 2006.

83 *Asahi Shimbun*, October 12, 2006.

84 It is one of the top three positions in the LDP except for the position of the party president.

85 *Kyodo News*, October 15, 2006, online edition.

86 *Reuters*, October 16, 2006, online edition.

87 *Shimbun Akahata*, October 19, 2006.

the major dailies in Japan, published an opinion by Susumu Nishibe, a well-known conservative scholar, that called for more discussion on Japan's option of going nuclear.[88]

While Japan strengthened its tough stance toward North Korea after its nuclear test, the U.S. moderated its stance. The U.S. had an unofficial meeting with North Korea in Beijing on October 31. It agreed with the DPRK to discuss the issue of the U.S. financial sanctions against the DPRK at the time of the next session of the six-party talks and succeeded in gaining North Korea's promise to return to the talks. Apparently annoyed by Japan's hostile stance, on November 4, North Korean Foreign Ministry maintained that it would be desirable for Japan not to be a part of the multilateral talks.[89] However, Japan's policy toward the North showed few signs of change.[90]

On November 14, 2006, the ruling coalition parties, the LDP and the Komei Party, formed a project team to discuss the introduction into the Diet of a draft of the space law to enable military use of space that would facilitate Japan's acquisition of a better satellite surveillance capability. On December 14, the project team decided to introduce the draft into the Diet in February 2007.[91]

Japan also took new actions on the abduction issue, which further soured its relations with North Korea. On November 2, 2006, Japan issued a warrant for arresting a North Korean agent for abducting a Japanese mother and daughter and demanded that the DPRK should hand him over. On November 20, 2006, Japan recognized Kyoko Matsumoto as the seventeenth Japanese citizen abducted by the DPRK and demanded her return from the DPRK. Thus, prior to the resumption of the six-party talks, there was a strong tension between Japan and the DPRK.

The Second Session of the Fifth Round of the Six-Party Talks

Thirteen months after the first session, the second session of the fifth round of the six-party talks began on December 18, 2006 and lasted for five days. On the sideline of the talks, the DPRK and the U.S. had bilateral talks on the U.S. financial sanctions on the DPRK. The second session ended without much progress because the DPRK refused to discuss concrete steps for implementing the joint statement of the fourth round of the six-party talks unless the U.S. lifted its financial sanctions. Reportedly, the U.S. presented to North Korea a multi-stage roadmap for North Korea's nuclear dismantlement, in which it clarified what the North Korea could get in return for its actions from the U.S. and other participants of the six-party talks.[92] Yet, the U.S.

---

88 Susumu Nishibe, "Kakubuso rongi no hirogari wo kitai suru," *Sankei Shimbun*, October 7, 2006, p.11.

89 *Nihon Keizai Shimbun*, November 4, 2006.

90 *Sankei Shimbun*, November 14, 2006.

91 *Asahi Shimbun*, December 14, 2006.

92 *Mainichi Shimbun*, December 20, 2006.

maintained that the nuclear issue and the sanction issue were separate and unrelated,[93] and then refused to lift the financial sanctions.

Japan did not play a constructive role in this session. On the first day, the Japanese envoy, Kenichiro Sasae, stressed that the abduction issue was the most important diplomatic issue to the Abe administration and that its diplomatic normalization with DPRK was conditional on the resolution of this issue.[94] Then, during its bilateral talks with the U.S., the PRC, the ROK, and Russia, Japan emphasized their understanding and cooperation on the issue.[95] Apparently to stress the primary importance of the abduction issue to the Abe administration, Sasae referred first to the abduction issue, while its reference order was nuclear, missile, and abduction in previous sessions.[96]

As of this writing, it is unclear whether Japan made any new proposals for the resolution of the nuclear issue. The government and the Ministry of Foreign Affairs stress that the Japanese envoy reiterated the importance of the abduction issue at the session, but they say little about what Japan did for the resolution of the nuclear issue. It is unclear whether Japan kept its previous stance that economic assistance to the DPRK would be conditional on a comprehensive resolution of the abduction, nuclear, and missile issues. What is clearly known is that Japan was the only country that did not have any official bilateral talks with the DPRK during the session.[97] It seems that all it had was a brief unofficial contact during a meal. It should be noted that no official dialogue has been held since February 2006. Sasae criticized the DPRK, saying "[The session] produced no tangible results because the DPRK was obsessed with the U.S. financial sanctions that are unrelated to the nuclear issue."[98] He did not make a self-evaluation concerning why Japan could not ease the tension between the DPRK and the U.S. and whether its policy stance toward the DPRK was appropriate.

Factors behind Japan's Response to the Second North Korean Nuclear Crisis

The preceding section examined Japan's response to the recent nuclear crisis, which began after suspicion of North Korea's uranium enrichment surfaced in October 2002. North Korea's nuclear development and the prospect of it being armed with

93 Stated by Assistant Secretary of State Christopher Hill and Secretary of State Condoleezza Rice. *Yonhap News*, December 22, 2006, online edition.

94 Gaimusho, *Dai5kai rokushakaigo dai2ji kaigo (gaiyo to tenbo)*, December 2006, online at <http://www.mofa.go.jp/mofaj/area/n_korea/6kaigo/6kaigo5_2gt.html>.

95 Ibid. Also, mentioned by the Chief Cabinet Secretary at a press conference held on December 18.

96 *Joongang Ilbo*, December 19, 2006, online Japanese edition.

97 After the dinner with the other delegates on December 17, Kenichiro Sasae told reporters that he had not requested his North Korean counterpart to hold bilateral talks during the session. *Jiji Tsushin*, December 17, 2006, online edition. Japan was uneager to make such a request; *Yomiuri Shimbun*, December 20, 2006; At a press conference on December 22, Sasae told that DPRK did not show any interest in talks with Japan. *Jiji Tsushin*, December 22, 2006, online edition.

98 *Mainichi Shimbun*, December 22, 2006, online edition.

nuclear missiles raised a serious security concern in Japan. Thus, Japan had a very strong reason to play an active role in stopping the DPRK's nuclear activities. Yet, as the preceding section showed, Japan's role at the six-party talks was not very constructive. Its stance toward North Korea has been generally inflexible, uncompromising, and confrontational.

Japan, along with the U.S., the ROK, the PRC, and Russia, stated its willingness to provide energy assistance to the DPRK in the joint statement adopted at the fourth round of the six-party talks. Yet, Japan has not made it clear whether it would take part in multilateral energy assistance to the DPRK if there was no significant progress on the abduction issue. Given its North Korea's expressed willingness to abandon its nuclear program in return for economic and security concessions, North Korea may actually move toward denuclearization if it is given sufficient concessions and a concrete idea of what it can get at which stages of the denuclearization.[99] However, Japan has been reluctant to give large economic concessions, while the U.S. has been unwilling to give enough economic and security concessions.

Why has Japan taken this kind of policy toward the DPRK? Three reasons can be conceived: the U.S. influence, the influence of the abduction issue, and the influence of hardliners and defense-related industries.

The U.S. Influence

Needless to say, Japan is heavily dependent on the U.S., both economically and militarily. This dependence has worked as a strong incentive for Japan to follow U.S. foreign policy, as can be seen from its past actions such as its active cooperation with the U.S. in its "war on terrorism" and the U.S. policy toward Iraq. For that matter, Japan also supported the Clinton administration's policy toward the DPRK by making a major financial contribution to KEDO's project of constructing two LWRs in the DPRK. Thus, it would be reasonable to assume that Japan's response to the second nuclear crisis has been strongly affected by U.S. policy toward the DPRK under the Bush administration. However, the U.S. influence does not seem to be the only determining factor.

The Influence of the Abduction Issue

As can be seen from Japan's stance at the six-party talks, the abduction issue has strongly affected Japan's policy toward North Korea's nuclear issue. Japan has given the highest priority to the resolution of the abduction issue. Tokyo's serious concern for the abduction issue is probably due to the strong public sentiment against the DPRK for abducting Japanese citizens and the people's strong desire to resolve this issue as soon as possible. It seems that the public is more concerned about the abduction issue than the nuclear issue. Tokyo would suffer harsh public criticism for

99 Some leading Japanese scholars hold the view that DPRK would not give up its nuclear weapons. Akihiko Tanaka, "Kitaidekinai Kitachosen no kaku haizetsu," pp. 38-43; Hideshi Takesada, "Kitachosen kakumondai no yukue," *Anzenhosho* 101 (January 2006), online at <http://park20.wakwak.com/~kokubou/books-101-3.html>.

giving significant economic concessions for North Korea's denuclearization without settling the abduction issue.

However, the supremacy of the abduction issue is partly a political creation. The government and key political figures such as Shinzo Abe, the former Chief Cabinet Secretary and the incumbent Prime Minister, played an important role in promoting the growth of the strong anti-North Korean sentiment and the increase in the salience of the abduction issue.[100] In this sense, Japan's policy toward North Korea has been constrained by its own creation. In contrast, the government and key political figures have played a very limited role in promoting public understanding about the importance of the nuclear issue.

The Influence of Hardliners and Defense-related Businesses

Another reason for Japan's hardline policy toward the DPRK over the nuclear, missile, and abduction issues can be found in the strong desires of its hardliners and defense-related businesses to pursue their objectives such as: (1) easing legal constraints on Japan's military activities, including greater use of space for military purposes, weapons export, and the expansion of its military power as well as the geographical scope of its military activities; and (2) securing long-term and lucrative defense-related business opportunities. The economic interests of defense-related businesses, as well as the conservative, nationalistic thoughts of hardliners, have strongly promoted the move to liberalize Japan's military activities and make Japan a "normal" state. For instance, Japan's major business federation, Keizai Dantai Rengokai [Keidanren, Japan Federation of Economic Organizations], and its successor, Nihon Keizaidantai Rengokai [Nippon Keidanren, Japan Business Federation], strongly supported the satellite project, the missile defense project, the liberalization of military exports, and the repeal of the Diet Resolution of the peaceful use of the space as ways to promote the growth of their business opportunities.[101] It would be reasonable to posit that hardliners and defense-related businesses expected that they could exploit Japan's tension with the DPRK to pursue their interests and therefore supported a hardline policy toward the DPRK. In fact, as shown above, Japan's hardline policy toward North Korea has increased the tension between Japan and North Korea, and the increased tension, in turn, has facilitated the liberalization of Japan's military activities and its financial commitment to big, long-term defense projects, such as the IGS project and the BMD project.

---

100 Japan's media has also played the same role.

101 Keizaidantai Rengokai, *Jiki chuki boeiryoku seibi keikaku ni tsuite no teigen* [A Proposal on the Next Midterm Defense Program] (Tokyo: Keidanren, September 2000), online at <http://www.keidanren.or.jp/japanese/policy/2000/045.html>; The U.S.-Japan Industry Forum for Security Cooperation (IFSEC), *IFSEC Joint Report: Revised U.S.-Japan Statement of Mutual Interests* (Tokyo: IFSEC, December 2002), online at <http://www.keidanren.or.jp/japanese/policy/2003/005e.html>; Nihon Keizaidantai Rengokai, *Kongo no boeiryoku seibi no arikata ni tsuite* [About the Future Course of the Military Buildup] (Tokyo: Keidanren, July 2004), online at <http://www.keidanren.or.jp/japanese/policy/2004/063.html>.

Future Prospects

The future direction of Japan's policy stance toward North Korea's nuclear development will depend largely on the U.S.'s policy choice. On the one hand, if the U.S. changes its uncompromising policy stance toward the DPRK and becomes willing to give major concessions and urges Japan to follow suit, then North Korea would become more willing to resolve the abduction issue in order to elicit major economic concessions from Japan. Also, such a soft-line policy of the U.S. would strengthen the voices of soft-liners in Japan and would reduce the criticism of a softer policy toward the DPRK from the public that has been supportive of an uncompromising stance to gain concessions on the abduction issue. Then, it would become difficult for the hardliners and defense-related industries to promote a hard-line policy. That would increase the possibility of Japan's adoption of a compromising stance toward the DPRK. If the U.S. and Japan join the ROK in pursuing a conciliatory policy toward the DPRK, then they can collectively offer strong incentives for the DPRK to abandon its nuclear ambition and thereby can significantly increase the chance of an early resolution of the second North Korean nuclear crisis.

On the other hand, if the U.S. maintains its uncompromising, hard-line policy toward the DPRK, it is doubtful that Japan will dare to give major economic concessions to the North, considering the strong U.S. economic and military influence over Japan. Also, the U.S. hard-line policy would make it unlikely for North Korea to give Japan major concessions on the abduction issue. As a result, the public animosity toward the DPRK would likely remain strong. Then, the soft-liners would have difficulty in promoting a conciliatory policy toward the DPRK, while the hardliners and defense-related businesses would find it relatively easy to prolong Japan's present hard-line policy and thereby would succeed in further liberalizing Japan's military activities and expanding defense-related business opportunities. Japan's greater military activities and military buildup, in turn, may well trigger a new arms race in East Asia and destabilize the region. Thus, not only the resolution of the North Korean nuclear crisis but also the military future of Japan and East Asia as a whole will depend strongly on U.S. policy choice.

Chapter 7

Russia and the North Korean Nuclear Crisis

Seung-Ho Joo

The second crisis over North Korea's nuclear issue erupted in the winter of 2002-03 after the U.S. accused the North of secretly and illicitly pursuing a uranium-based nuclear weapons program. Tensions between the U.S. and the Democratic People's Republic of Korea (DPRK) heightened and a direct military confrontation on the Korean peninsula loomed large. In an effort to resolve the crisis through dialogue, the six-nation talks, involving the U.S., China, Russia, Japan, and the two Koreas, convened in August 2003 in Beijing. Russia thus came to participate in multinational talks on North Korea's nuclear problem for the first time. The six-party talks convened intermittently in the following three and half a years. The North Korean nuclear talks frequently stalled, failing to bridge the wide gap between the U.S. and North Korea. At the end of the fourth round of the six-party talks in August 2005, the six nations reached the first significant agreement on the Statement of Principles in which North Korea agreed to abandon its nuclear weapons and nuclear programs in exchange for economic, security, and political compensations. The U.S. imposition of financial sanctions on the DPRK in the same month strained U.S.-North Korean relations and put a curb on the six-party talks. In February 2007, the six-party talks produced a landmark accord, "Initial Actions for the Implementation of the Joint Statement," which could lead to the ultimate denuclearization of North Korea.

After Vladimir Putin came to power in 2000, Russia resumed normal relations with North Korea and has been energetically involved in Korean affairs in an effort to safeguard its political, economic, and security interests. In the six-party negotiations, Russia has played a constructive role for Korean denuclearization and Northeast Asian security and prosperity. This chapter focuses on Russia's role in the on-going six-party talks on North Korea's nuclear problem. More specifically, this research raises the following questions and tries to answer them: What are Russian interests and policies toward the Korean peninsula? What role has Russia played at the six-party talks? How has Russia sought to contribute to the denuclearization of the Korean peninsula?

Russia and the Korean Question

At the end of Yeltin's presidency, Russia was in disarray. The economy was in shambles, political powers divided, and foreign policy adrift. Under Putin,

Russia's economy has been steadily growing, its political powers are centralized, and its foreign policy is firmly anchored. Putin's foreign policy is characterized by pragmatism, balance, and realism. President Putin has pursued the twin goals of economic development and national security. Russian foreign policy has been geared toward creating favorable external conditions for economic development (marketization and global integration). In parallel, Russia has pursued the goal of creating a multipolar world by opposing the current U.S.-dominated, unipolar power structure and reestablishing itself as a great power of Eurasia.[1] There are inherent contradictions between these two sets of goals since Russia's economic development and global integration will not be likely without U.S. support. The reality is that Russia cannot completely antagonize the U.S. if it wants to modernize its economic system and integrate it into the world economy. And the fact is that the interests of the U.S. and Russia do not always coincide. In areas of common interest, such as anti-terrorism, destruction (WMDs), and regional stability, Russia joins hands with the U.S. to cope with common threats and problems. In issue areas where they do not see eye to eye, Russia is increasingly assertive of its interests and viewpoint.

The Russian Far East has a special meaning in Putin's foreign policy due to its critical location (the link between East Asia and Europe) and abundant resources (gas, oil, and other mineral resources). Like his predecessors (Gorbachev and Yeltsin), Putin believes that the development of the Far East and Russia's integration into the economic and security structures of the Asia-Pacific will provide the momentum for Russia's overall economic development and guarantee Russia's security in the region. In this context, then Deputy Foreign Minister Grigori Karasin in February 2000 enunciated the three main goals of Russian foreign policy in the Far East in terms of stability, border security, and cooperation, "First, Moscow seeks 'maximum participation in international security structures' to help ensure 'stability and predictability' in that region. Second, it aims for the security of its borders and the introduction of long-term confidence-building measures. And third, it wants to establish political and economic relations with all countries of the region that would help promote the development of Russia's Far East."[2]

Korea takes up a special place in Russian foreign policy due to its geo-strategic location, North Korea's nuclear crisis, and Russia's enormous potential for economic cooperation with the two Korea's, especially in the fields of energy and transportation. Russia's goals and interests vis-à-vis the Korean peninsula may be summarized as the following. First, Russia is opposed to any power dominating the Korean peninsula. Since Korea is geo-strategically important to the Russian Far East, Russia considers Korea's domination by one alien power a direct and grave threat to its security. Russia will thus seek to minimize U.S., Chinese, or Japanese influence over the Korean peninsula. Second, Russia advocates a multinational

---

1 For the principles, goals, and priorities of Russian foreign policy, see Russian Foreign Policy Concept presented on July 10, 2000 by the Russian Foreign Ministry. The full text of the document is available at "The Foreign Policy Concept of the Russian Federation," *Nezavisimaya Gazeta*, July 11, 2000, pp. 1, 6, in *Current Digest of the Post-Soviet Press*, vol. 52, no. 17 (2000), pp. 13-15 & vol. 52, no. 29 (2000), pp. 6-8.

2 ITAR-TASS, February 7, 2000.

security mechanism as a vehicle to the final settlement of the Korean question. Given Korea's geo-strategic location, the major powers (the U.S., China, Japan, and Russia) surrounding the Korean peninsula have inherent interests in Korea's future. Cooperation and coordination among the major powers is thus a prerequisite for a lasting peace and security on the Korean peninsula. Based on this rationale, Russia has long advocated a multinational dialogue to deal with the Korean question. The two Koreas, Russian pundits point out, should play the leading role in Korean peace process, and the major powers should play a complementary role.[3] Third, Russia wants to maintain peace and stability on the Korean peninsula. Russia does not want to see another Korean war since it would inevitably disrupt Russia's efforts to develop the Far East and implement modernization at home. Furthermore, the destruction of Korean nuclear reactors and the influx of Korean refugees into Russia, in case of another Korean conflict, would threaten the security of the Russian Far East.[4] Fourth, Russia favors a gradual and incremental approach to settling the Korean question. The two Koreas should pursue a long-term peaceful coexistence before they achieve unification, and neither the U.S., nor South Korea should attempt to change North Korea's behavior or seek North Korea's collapse. Korean unification should be achieved through peaceful means and the two Koreas should negotiate a peaceful unification on an equal footing. Russia prefers to see a unified Korean state that is friendly or neutral. Russians believe that a neutral and unified Korea will be in Russia's national interest.[5] Russia wants to participate along with other powers in the Korean unification process on an equal footing in order to protect its interests. Fifth, Russia is likely to oppose U.S. military presence in the unified Korea. The unified Korea as a military ally of the U.S. would mean for Russia an "Asian version of NATO's eastward expansion" or a "forward military base on Russia's doorstep."[6] But Russia's attitudes toward U.S. military presence in Korea and a U.S.-Korean alliance after Korean unification will depend to a great extent on the Northeast Asian power structure that will take shape in the future.

Putin's Russia maintains an even-handed relationship with the two Koreas. Departing from the pro-Seoul, lopsided policy of his predecessors (Gorbachev and Yeltsin), Putin has sought to maintain a balanced relationship with the two divided Korean states. The new foreign policy concept of 2000 cited an active involvement in the settlement of the Korean question and the maintenance of a balanced relationship with both Koreas as the main goals of Russia's Korea policy. Russia's rapprochement with the DPRK in 2000 was a natural corollary of Putin's

---

3 G. Toloraya, "Security and Confidence Building in Korean Peninsula: A Russian Point of View," LNCV-Korean Peninsula: Enhancing Stability and International Dialogue, June 1-2, 2000, Roma, online at <http://lxmi.mi.infn.it/~landnet/corea/proc/013.pdf>.

4 See Vadim Tkachenko, "A Russian View on Korean Security after the North-South Summit," *The Korean Journal of Defense Analysis*, vol. 12, no. 2 (winter 2000), p. 28.

5 See Vadim P. Tkachenko, *Koreiskii Poluostrov i Interesy Rossii [The Korean Peninsula and Russian Interests]* (Moscow: Vostochnaia Literatura, 2000); Vladimir F. Li, Russiia I Koreia v Geopolitike Evroziiskov *[Russia and Korea in the Eurasian Geopolitics]* (Moscow, 2000).

6 Vadim Tkachenko,"A Russian View on Korean Security after the North-South Summit," p. 31.

pragmatic foreign policy line. From the Russian viewpoint, the prerequisite to playing a central role in the Korean peace process is to maintain friendly ties with both Koreas. Accordingly, Russia achieved rapprochement with North Korea in 2000 and reestablished a bilateral, good-neighborly relationship in multiple spheres, while forging a "comprehensive strategic partnership" with the Republic of Korea (ROK or South Korea). Russia is even-handed between the two Korean states in political spheres only. Russia inevitably leans toward South Korea in almost all other spheres due to the fact that as a partner South Korea has far more to offer to Russia than North Korea does.

The Second North Korean Nuclear Crisis

The 1994 U.S.-DPRK Agreed Framework on North Korea's nuclear issue collapsed and a second crisis over the North's nuclear ambitions flared up in October 2002. During U.S. special envoy James Kelly's visit to Pyongyang in early October 2002, Kang Sok-Ju, First Deputy Foreign Minister of North Korea, allegedly admitted to a secret nuclear program based on highly enriched uranium (HEU).[7] In December, the U.S. halted heavy fuel oil shipments to North Korea in violation of the Agreed Framework. In response, North Korea declared the 1994 Agreed Framework null and void, removed monitoring devices and expelled the International Atomic Energy Agency (IAEA) inspectors. North Korea then resumed its nuclear program, which had been mothballed since October 1994, by reopening a sealed plutonium-reprocessing plant at Yongbyun. On January 10, North Korea announced its decision to withdraw from the Nuclear Non-proliferation Treaty (NPT). It appeared that the U.S. and North Korea were on a collision course and an armed clash between the two nations was imminent. In the midst of deepening crisis, the North insisted that the nuclear problem be resolved through direct U.S.-DPRK talks. In contrast, the U.S. maintained that the crisis should be settled through multilateral talks.

Until March 2003, the U.S. adhered to military pressure tactics toward the DPRK and opposed Russia's participation in a multilateral negotiation format over North Korea's nuclear issue. After the initial efforts to cope with the crisis in the trilateral setting (the U.S., China, and the DPRK) failed in April, the U.S. wanted to hold five-party talks, including the U.S., China, the DPRK, the ROK, and Japan. China then dispatched Vice-Foreign Minister Dai Bingguo to Pyongyang to convey the U.S. intention. At his meeting with Kim Jong-Il, Dai urged him to accept five-way talks, but Kim rejected it and counter-proposed the six-party format. According to *Sankei Shinbun*, in late July 2003, Kim Jong-Il telephoned Russian President Putin and requested that Russia participate in six-party talks as the host. Reportedly Putin agreed to Russia's joining the talks, but declined to host them on the grounds that Beijing was the deserving host in view of its on-going, sincere peace efforts between the U.S. and the DPRK.[8]

---

7 'Former US Envoy: US Did Not Confront DPRK with Uranium Enrichment Proof," *Yonhap*, November 20, 2003.

8 Tadashi Ito, "PRC Source Cited on Putin Rejecting Kim Chong-il Request to Host Talks in Russia," *Sankei Shimbun*, September 9, 2003.

Until the first round of the six-party talks began on August 27, 2003 in Beijing, Russia had been excluded from multinational deliberations on North Korea's nuclear question. For years, Russia had advocated a six-party format as a vehicle for peace and security in and around Korea. However, neither the U.S. nor the DPRK paid much attention to this proposal. In 1994 Russia complained bitterly about its exclusion from the negotiating process leading to the October 1994 U.S.-DPRK Framework Agreement, and in 1996 it expressed strong regrets over its exclusion from the four-party—the U.S., China, North and South Korea—Korean peace talks. This time, Russia quietly but effectively pushed for its inclusion in multinational talks over the North Korean problem, and such diplomatic efforts paid off.

The idea of six-party talks (or multilateral talks including "all interested parties") to discuss Korean issues originated from the Russian side. For years, Russia had consistently advocated a multinational format (6-party, 8-party or 10-party talks) to discuss issues relating to Korean peace and security, while maintaining that the U.S. alone cannot untie the "Korean knot."[9] Russia's call for multinational talks on the Korean issue fell on deaf ears until July 2003 when the U.S. and North Korea accepted the six-party format. At the time, the Bush administration agreed to Russia's inclusion in the six-party talks in the hopes of enlisting Russia's support to force the DPRK into dismantling its nuclear programs. The DPRK, in contrast, wanted Russia's inclusion in order to offset the U.S. military threat and inflexible stance at the negotiating table. The convening of the six-way talks was certainly a diplomatic triumph for Russia as it finally gained a foothold in a multinational forum to deliberate on peace and security in Korea and Northeast Asia.

Russia at the Six-Party Negotiation Process

The first session of the six-party talks convened on August 27-29, 2003 in Beijing, but failed to produce any substantive results. The talks, however, established the objective of a nuclear-free Korean peninsula and set in motion the Korean peace process. During the talks, the DPRK confirmed its overall objective of the nuclear-free Korean peninsula and the U.S. promised not to attack, threaten, or invade the DPRK. At the Beijing talks, the U.S. and the DPRK failed to reconcile their diametrically opposed positions.

After a six-month hiatus, the second round of six-party talks was held on February 26-28, 2004 in Beijing. No breakthrough was made at this session as the U.S. and North Korea stuck to their previous positions. As before, the U.S. refused to provide the DPRK with a written security guarantee or economic aid before the DPRK implements "complete, verifiable and irreversible" dismantlement (CVID) of its nuclear programs. Meanwhile, the DPRK insisted on a package solution that North Korea's renunciation of the nuclear programs and U.S. compensations to the DPRK should proceed simultaneously. During the talks, North Korea offered to freeze its nuclear programs in exchange for aid, and South Korea offered to provide energy

9 As early as 1992, Russian President Boris Yeltsin proposed a multinational negotiating mechanism in Northeast Asia. In fall 2002, Russian Deputy Foreign Minister Alexander Losyukov reiterated its call for six-party talks to discuss North Korea's nuclear issue.

aid if the freeze was the first step to dismantlement. China and Russia offered to provide aid along with South Korea.[10] At this session, the six nations agreed to form smaller working groups that would handle more substantial and technical aspects of the nuclear dispute.

The first working-group meeting of the six-party talks was held on May 12-14, 2004 in Beijing. The open-ended working-group talks were intended to lay the groundwork for the third session of the six-party talks scheduled to be held by the end of June. At the meeting, the DPRK once again put forth the "reward for freeze" proposal. The U.S. in turn reiterated its previous position that the DPRK should accept CVID before talks on compensations begin.[11] During the talks, China and Russia focused on "compensation for freezing."[12] The impasse over North Korea's nuclear issue continued.

As Russian Deputy Foreign Minister Alexander Losyukov pointed out, the "working group" session was Russia's idea.[13] In fact, the working-group idea dates back to 1997 when Russia proposed a 10-party multilateral conference—the DPRK, the ROK, the five permanent members of the UN Security Council, Japan, the UN Secretary General, and the General Director of the International Atomic Energy Agency (IAEA)—on the Korean question. This proposal included clauses on the formation of working groups to deal with specific issue areas and included the specific details of their operation.[14] Although the six-party conference differs from the Russian-proposed 10-party conference in terms of context and preconditions, the idea of forming a working-group was adopted at the second full session of the six-party talks for its pragmatic application.

In July-September 2005, the fourth round of six-party talks was held in Beijing. On September 19, the first breakthrough was reached when the six parties agreed on a "statement of principles," which stipulated Pyongyang's dismantling of all its nuclear programs in exchange for diplomatic recognition, security guarantee, and economic aid.[15] Some parts of the agreement were vague and ambiguous. Still, the September 19 joint statement was an important breakthrough that could pave the

10 Joseph Kahn, "U.S. and North Korea Agree to More Talks," *New York Times*, February 29, 2004.

11 Jack Kim and John Ruwitch, "North Korea Crisis Talks End on 'Wholesome' Note," *Reuter*, May 14, 2004.

12 "Allies Trying to Persuade NK on Nuke Programs," *Korea Times*, May 13, 2004.

13 Valery Agarkov, "Russia: Losyukov May Head Russia Delegation at North Korean Nuclear Talks," *ITAR-TASS*, February 3, 2004.

14 Valentin Moiseev, then Deputy Director of the First Asian Department of the Russian Foreign Ministry, published an article in the May-June, 1997 issue of *International Affairs* (the journal published by the Russian Foreign Ministry) that, for the first time, included concrete details of the Russian proposal. For further details, see Valentin Moiseev, "On the Korean Settlement," *International Affairs* (Moscow), vol. 43, no. 3 (1997), pp. 68-72.

15 For full text of the agreement, see "Full Text of Joint Statement of 6-way Talks," *Japan Economic Newswire*, September 19, 2005; "Text of Joint Statement from Nuclear Talks," *The New York Times*, September 19, 2005.

way to a peaceful resolution of North Korea's nuclear crisis.[16] The statement was designed to serve as the basis for further talks on the timing of North Korea's nuclear dismantlement and corresponding provisions of economic aid and diplomatic relations and other rewards for the North.

Alexander Alexeyev, then Russian chief negotiator at the six-party talks, hailed the joint statement of principles as profoundly significant. The key issue at the fourth round of talks was North Korea's demand for its right to peaceful uses of nuclear energy. The North insisted that it be allowed to have a peaceful nuclear program and a light water reactor (LWR) be provided in compensation for abandoning its nuclear programs. The U.S. maintained that North Korea should abandon all nuclear programs, including civilian nuclear power programs. The U.S. and the DPRK in the end agreed to a compromise. The U.S. agreed to accept North Korea right to peaceful uses of nuclear energy and to discuss "at an appropriate time" the supply of LWR to the North. The DPRK, in turn, agreed to return to the NPT and IAEA "at an early date" and to dismantle all its nuclear programs and nuclear weapons. The timing and sequence of North Korea's denuclearization and the other nations' supply of an LWR to the North were to be resolved in subsequent negotiations.

Russia held that North Korea should have the right to peaceful nuclear energy programs in light of its acute energy shortages but an LWR reactor should be supplied to the North only after it returns to the NPT and accepts IAEA inspections.[17] In a speech given at Stanford University on September 21, 2005, Russian Foreign Minister Sergey Lavrov said that LWRs should be provided to North Korea only after it abandons its nuclear weapons and related programs: "We think North Korea's nuclear weapons should be scrapped first and then the issue of the LWRs decided afterwards."[18] Deputy Foreign Minister Alexeyev stated that the next round of talks should focus on a "road map" to specifically implement the agreed elements and carry out North Korea's denuclearization based on the principle of synchronization: "It is necessary to first of all determine the sequence, nature and volume of the steps to be taken by the parties in conformity with the synchronization principle in order to give a start to the process of nuclear disarmament on the peninsula."[19] Russian Ambassador to Seoul, Glev A. Ivashentsov said in an interview in November that the provision of LWRs to North Korea and the North's dismantling of nuclear weapons should take place simultaneously, "Here you see an issue of mutual mistrust. The only

16 For a detailed analysis of the six-party talks, see Tae-Hwan Kwak, "Resolving the North Korean Nuclear Issue Through the Six-Party Process: A Creative Formula," paper presented at the 47th International Studies Association Annual Meeting, San Diego, CA, March 22-25, 2006.

17 See Deputy Foreign Minister Alexander Alexeyev's statements in "Russian Delegation: North Korean Nuclear Programs May Have Double Use," *ITAR-TASS,* September 16, 2005; "Xinhua Roundup: Positions of Delegations To Six-Party Talks," *Xinhua,* September 15, 2005.

18 "ROK's Yonhap: Russia Demands N.K. Scrap Nuclear Weapons Before Getting Lwrs," Yonhap, September 21, 2005.

19 "Russian Official: N Korea Nuclear Talks Parties Need To Bring Stances Closer," *ITAR-TASS*, October 20, 2005.

solution, therefore, is all the steps synchronize."[20] Russia expressed its willingness to provide energy assistance to North Korea and build a nuclear reactor in North Korea but only after the DPRK rejoins the NPT and IAEA. It is, however, unlikely that the DPRK will receive energy assistance from Russia before U.S-DPRK relations substantially improve or the six-party talks make serious progress. It is also unlikely that a Russian nuclear reactor will be built on North Korean soil since South Korea, which is most likely to provide the bulk of the construction funds, would insist on a nuclear reactor of its own model.

On September 15, 2005, the U.S. Treasury Department designated the Banco Delta Asia (BDA) in Macau as a financial institution of "primary money-laundering concern" under Section 311 of the USA Patriot Act.[21] The Treasury claimed that for 20 years BDA was a "pawn for the North Korean government to engage in corrupt financial activities through Macau," providing financial services such as circulating counterfeit money and laundering of illicit money to North Korean firms linked to the smuggling of drugs, fake tobacco and precious metals.[22] The U.S. Treasury Department also alleged that BDA facilitated multimillion dollar wire transfers for North Korean firms.[23] In accordance with the U.S. Patriot Act, the Treasury Secretary prohibited U.S. financial institutions from conducting any business transactions with BDA in the U.S. for or on behalf of the bank. Subsequently, BDA froze North Korea's 50 accounts with the bank. Although the amount frozen as a result of the sanctions was only $24 million, the move had a serious impact on DPRK's financial capability because it did not have many windows to the international financial market and it received international aid and conducted international financial transactions through the bank.

During the fifth round of the six-party talks held on November 9-11, 2005, the parties sought to find ways to implement the provisions of the joint statement based on the principle of "commitment for commitment, action for action," but no progress was made as the North refused to return to the negotiating table unless the U.S. lifts the financial restrictions first. Pyongyang denied all charges of illicit financial activities. Russian Ambassador to Seoul Ivashentsov stated that his country did not have substantial evidence on the U.S. charge and demanded concrete evidence: "Russia has not received any concrete evidence [of North Korea's alleged counterfeiting]. There is rumor-level talk on the issue." He further reiterated Russia's

---

20 "Nk Nuke Scrapping, Lwr Provision Should Take Place Simultaneously," *Yonhap*, November 1, 2005.

21 Section 311 of the USA Patriot Act authorizes the Secretary of the Treasury to require domestic financial institutions and agencies to take special measures "if the Secretary finds that reasonable grounds exist for concluding" that a foreign financial institution is of primary money laundering concern. See USA PATRIOT Act of 2001, available online at <http://www.ratical.org/ratville/CAH/Section301.html#311>.

22 For details, see the first statement issued by the US Treasury Department at <http://www.ustreas.gov/press/releases/js2720.htm>; For see money laundering at <http://usinfo.state.gov/ei/economic_issues/terrorist_financing/money_laundering.html>. See also "Breaking the bank: A bank run in Macau.," *The Economist*, September 24, 2005, p. 90.

23 "Macau-based Chinese Bank Named Money Launderer For North Korea," *Yonhap*, September 15, 2005.

position that sanctions on North Korea would not solve any problem and a solution would be possible only through dialogue and cooperation.[24] In the meantime, the Iranian nuclear crisis erupted in early 2006 and once again North Korea's nuclear issue was sidelined.

North Korea's Missile and Nuclear Tests

On July 5, 2006, North Korea decided to up the ante by test-firing seven ballistic missiles, including a Taepodong-2 long-range missile, which has the potential of reaching the U.S. west coast. The Taepodong-2 test failed as it crashed into the ocean after 35-40 seconds of its launch. North Korea's multiple test launches of missiles invited immediate, angry reactions throughout the world. The UN Security Council (UNSC) adopted Resolution 1695 on July 15, which condemned the tests and paved the road for international economic sanctions on North Korea.[25] Following the missile tests, South Korea and Japan unilaterally took punitive actions against North Korea. It appears that North Korea's missile tests were motivated by an attempt to gain leverage at the six-party talks and to attract U.S. attention.

The missile test-firings came as a surprise to Russia as North Korea failed to provide it with notification in advance.[26] In response to the missile launches, Putin expressed his disappointment at North Korea's inappropriate conduct.[27] Russian officials, however, warned against any emotional reactions and stated that other states should persuade North Korea to return to the missile launch moratorium and the six-party talks rather than provoke the country on purpose.[28] Russia also maintained that the UN Security Council's response to the North Korean missile crisis should be calibrated and should not contain threats. Moscow, like China, did not veto the UNSC Resolution 1695 but ensured that the resolution did not include Chapter 7, which authorizes military action. Moscow wanted to demonstrate its displeasure at North Korea's missile launches but at the same time wanted to avoid the tense situation getting out of control. Head of the Russian Federal Space Agency (Roskosmos) Anatoly Perminov assessed from the results of the launch that North Korea's missile program was at the initial stage, "There were major errors in the

24 Lee Chi-dong, "Russia Urges US to Present Evidence on N Korean Counterfeiting," *Yonhap*, March 7, 2006.

25 The full text of UN Security Council resolution 1595 is available online at <http://www.un.org/News/Press/docs/2006/sc8778.doc.htm>.

26 "Russian Source Says Pyongyang Did Not Notify Moscow of Missile Launches," *Interfax*, July 5, 2006. North Korea did not notify China of its missile tests in advance either. "Chinese Source Says No DPRK Notification of Missile Tests," *Interfax*, July 6, 2006.

27 "Russia disappointed over North Korean missile launches," *Agentstvo Voyennykh Novostey*, July 6, 2006.

28 "Russian President Says North Korea Should Not Be Provoked 'on Purpose,'" *Yonhap*, July 13, 2006; "Russian minister advocates 'persuasion' to end Pyongyang's missile tests," *ITAR-TASS*, July 16, 2006.

preparation and carrying out of the launchings, as well as at the stage of the missiles flight."[29]

In a daring move, the DPRK carried out its first ever underground nuclear test on October 9, 2006, which sent out shock waves throughout the world. Russia and China had strongly urged North Korea not to go ahead with a nuclear test, but Kim Jong-Il ignored their pleas and demonstrated to the world his country's nuclear weapons capability through the test. Two hours prior to the nuclear test, the DPRK's deputy foreign minister informed the Russian Ambassador in Pyongyang of the planned test. North Korea, however, notified China of the test only 30 minutes in advance. China then conveyed the information to the U.S., Japanese, and South Korean governments.[30] North Korea's nuclear blast deeply disappointed and angered Russian and Chinese leaders, who decided not to veto a UNSC resolution against the recalcitrant state.

The North's nuclear test promptly led to the adoption of UNSC Resolution 1718 under Chapter 7, Article 41 of the UN Charter, which was aimed at penalizing North Korea for its nuclear test with diplomatic and economic sanctions.[31] UNSC Resolution 1718, adopted unanimously on October 14, prohibited the transfer to and from North Korea of the weapons of mass destruction (nuclear, chemical or biological weapons), their means of delivery (ballistic missiles) and related materials. It also prohibited exports of luxury items to North Korea. The U.S. and Japan were particularly harsh and swift in condemning North Korea's reckless behavior and sponsoring UNSC Resolution 1718. All Security Council members, including China and Russia, agreed that "there needs to be some appropriate punishment" for North Korea's nuclear detonation. But South Korea, China, and Russia were opposed to any use of force and issued warnings against escalating the situation out of control.[32] The DPRK rejected the resolution and threatened "physical countermeasures" against any state that would try to enforce the U.N. sanctions.[33]

In a live Russian TV program on October 25, Putin commented on North Korea's recent nuclear test. While strongly disapproving of the nuclear blast, he warned against driving North Korea into a corner and urged all the parties to return to the six-party talks to resolve the crisis through dialogue. He also blamed the US for its heavy-handed attitudes at the negotiating table and driving North Korea into a dead-

29 "Russian Space Chief Says N. Korean Launches Highlight Technical Flaws," *ITAR-TASS,* July 6, 2006.

30 Tadashi Ito, "General Secretary Kim Succumbs to China's Pressure; DPRK Returns to Six-Party Talks in Fear of Expanded Sanctions," *Sankei Shimbun,* November 6, 2006; Tadashi Ito, "General Secretary Kim Succumbs to China's Pressure; DPRK Returns to Six-Party Talks in Fear of Expanded Sanctions," *Sankei Shimbun,* November 6, 2006.

31 For the full test of United Nations Security Council Resolution 1718, go to <http://www.nautilus.org/fora/security/0688UNSCStatement.html>.

32 John O'Neil and Choe Sang-Hun, "China Shows Willingness to Punish North Korea for Test," *New York Times*, October 10, 2006.

33 "DPRK Foreign Ministry Spokesman Totally Refutes UNSC 'Resolution,'" KCNA, October 17, 2006 available online at <http://www.kcna.co.jp/index-e.htm>.

end.[34] During the APEC forum held in Hanoi in November 2006, Russian First Deputy Foreign Minister Andrei Denisov clarified Russia's position on the UN sanctions against North Korea. The main purpose of the sanctions, he stated, is not to punish North Korea but to urge the country to return to the NPT and settle the North Korean nuclear crisis at the negotiating table: "Our purpose is not to punish somebody, but to achieve the main goal: observance of the non-proliferation regime and Korea's comeback to the Nuclear Weapons Non-Proliferation Treaty regime, and thereby to settle the nuclear problem on the Korean Peninsula."[35] Russian Ambassador to Seoul Ivashentsov reaffirmed in December 2006 Russia's policy of not recognizing North Korea as a nuclear power and reiterated its position that North Korea should have the right to peaceful uses of nuclear energy.[36]

According to a Russian report cited by the UN News Center, Russia pledged to implement the UNSC Resolution 1718, "Russia will fulfill United Nations' Security Council Resolution 1718 adopted on October 14, 2006. As a permanent member, Russia participated in developing the list of objects, materials, equipment, goods and technologies that are banned from being delivered to North Korea." The report further emphasized the need to settle the North Korean crisis politically: "At the same time, we think that the main goal of the resolution is to politically settle the North Korean nuclear issue, rather than to punish the country."[37]

The U.S. and Japan sought to apply the Proliferation Security Initiative (PSI) to North Korea. PSI is a US policy that is part of its global war on terror. If implemented, it will allow states to search and interdict North Korean ships and aircraft for contraband items (the weapons of mass destruction, missiles, and illegal weapons) on and over the high seas. Russia, China, and South Korea, however, have refused to fully participate in PSI.[38] The Bush administration's heavy-handed approach to North Korea proved counter-productive. Full scale economic or military sanctions against North Korea were most likely to be ineffective as long as South Korea and China do not want to push North Korea into a corner. China has argued that the problem of financial sanctions must be solved for the denuclearization of the Korean peninsula and that the U.S. should engage in bilateral talks with North Korea to resume the six-party talks. Russia also shares the international community's concern and anger over North Korea's missile and nuclear tests but, like China, is unwilling to pursue the goal of North Korea's regime change or collapse.

---

34 "Putin Says DPRK Nuclear Test Occurred Due to Dead-End Situation," *Russia OSC Report,* October 25, 2006.

35 "Russian Minister Comments on APEC Discussions on North Korea," *ITAR-TASS,* November 16, 2006.

36 "Russia Firmly Against North Korean Nuclear Status-ambassador," *ITAR-TASS,* December 20, 2006.

37 "Russia To Observe Ban On Deliveries Of Luxury Goods To North Korea," *Agentstvo Voyennykh Novostey,* December 13, 2006.

38 For an in-depth analysis of interdicting North Korean vessels and aircraft for WMD, see Mark J. Valencia, "Maritime Interdiction of North Korean WMD Trade: Who Will Do What?" *Policy Forum Online 06-98A: November 3rd, 2006,* available online at <http://www.nautilus.org/fora/security/0698Valencia.html>.

From Berlin to Beijing: A Breakthrough

After a 13-month hiatus, the six-party talks resumed on December 18-22, 2006. This time, the U.S. proposed a four-stage denuclearization plan to the DPRK: freeze, declaration, verification, and dismantling. North Korea, however, refused to discuss nuclear issues and focused instead on the financial sanctions issue. Washington and Pyongyang failed to narrow their differences over the financial sanctions on North Korea, and consequently the second phase of the fifth round of six-party talks failed to produce any substantive results. Still the two sides agreed to continue their talks on the financial sanctions in New York in January 2007.[39]

Assistant Secretary Christopher Hill and Vice Minister Kim Kye-gwan held three-day talks in Berlin on January 16-18, 2007 to discuss the nuclear and the BDA issues. This marked the first direct, bilateral talks between the U.S. and North Korea outside the six-party framework. At the Berlin talks, the U.S. and the DPRK agreed to hold parallel talks on the nuclear issue and U.S. financial restrictions issue. In an optimistic atmosphere, the fifth round of six-party talks reconvened on February 8-13, 2007 in Beijing. After five days' negotiations, the six nations produced a landmark accord which could lead to the ultimate denuclearization of North Korea.[40] The agreement laid out the first concrete steps to carry out the September 19, 2005 agreement, in which North Korea pledged to dismantle its nuclear programs in exchange for compensations. In this deal, North Korea agreed to begin disabling its nuclear facilities in exchange for food, fuel and other aid from the United States, China, South Korea and Russia.[41] The four countries agreed to provide about $400 million worth of aid in return for North Korea's commencement of the process of permanently dismantling its nuclear programs. Besides, the United States and Japan agreed to discuss normalizing relations with North Korea, and the United States agreed to begin the process of removing North Korea from its list of terror-sponsoring states, and lifting trade and financial sanctions. Under the agreement, North Korea would shut down its five nuclear facilities, including the 5 megawatt nuclear reactor in Yongbyon, within 60 days and South Korea and Russia would provide North Korea with 50,000 tons of fuel oil, or an equivalent value of economic or humanitarian aid. In return for disabling the reactor and declaring all its nuclear programs, the North was to receive another 950,000 tons of oil in stages. The accord stipulated that in the 60-day period the U.S. should begin to implement its promises to start bilateral discussions on various issues with the North. The six-nations also agreed to form and operate working groups in five issue areas.

---

39 "Fifth Round Of Six-nation Negotiations Over In Beijing," *ITAR-TASS*, December 22, 2006.

40 The full text of the February 13 Agreement is available online at <http://americancorners.or.kr/e-infousa/wwwh5668.html>. See also Edward Cody, "Tentative Nuclear Deal Struck With North Korea: Steps to Disarmament Drafted in 6-Party Talks," *Washington Post*, February 13, 2007; Jim Yardley, "North Korea to Close Reactor in Exchange for Aid," *New York Times*, February 13, 2007.

41 Japan refused to provide aid to North Korea along with the four countries and stated that bilateral issues between North Korea and Japan, including the abductions of Japanese citizens by the North, should be resolved first.

No doubt this accord provided an invaluable breakthrough but intractable issues remain and will have to be resolved in future negotiations. First of all, the agreement does not include issues relating to North Korea's nuclear weapons and nuclear weapons material. Second, the agreement does not mention North Korea's alleged secret HEU nuclear program. Third, the deal does not include a provision of light water reactors to North Korea. The February 13 agreement was made possible because both North Korea and the U.S. changed their initial positions and decided to make concessions for a compromise solution. North Korea no longer insisted on the US lifting the financial restrictions imposed on its country. The North was also willing to accommodate the U.S. demand for cracking down on illicit financial transactions and monetary earnings. The US promise at the Hill-Kim gathering in Berlin to unfreeze $11 million of the $24 million frozen at BDA in North Korean accounts appears to have been critical in facilitating the February nuclear deal. U.S. willingness to discuss the financial restrictions in a bilateral setting and its promise to partially lift the sanctions saved face for North Korea and signaled a change of US hostile policy toward the Kim Jong-Il regime.

Russian Efforts for Korea's Denuclearization

Russia has endeavored to bring about Korea's denuclearization in the six-party negotiation process and outside it. These efforts have been made both jointly, especially with China, and unilaterally.

It is noteworthy that Russia has been coordinating and consulting frequently and closely with China throughout the whole negotiation process. During the first round of the six-party talks, Moscow and Beijing delegations worked closely for the talks' fruitful conclusion. China was the first country with which Russia held diplomatic consultations in 2004. When Russian and Chinese diplomatic representatives met in Moscow on January 5, 2004, the North Korean nuclear issue topped the agenda.[42] Moscow and Beijing frequently discussed ways of settling the North Korean crisis peacefully. In this regard, Russian Ambassador to Beijing Igor Rogachev stated: "Sometimes we met almost every day, exchanged views, and compared our stances, in short, worked very closely together."[43] Russia also held a series of talks with South Korean and Japanese officials and had comprehensive discussions with the U.S. officials, including discussions in Washington in November 2003.[44] On the eve

42 Valery Agarkov and Yevgeny Tkachyov, "Russia, China Begin Consultations on North Korean Nuclear Problem," *ITAR TASS,* January 5, 2004.

43 Valery Agarkov, "Russia: Losyukov May Head Russia Delegation at North Korean Nuclear Talks," *ITAR-TASS,* February 3, 2004.

44 During his two-day stay in January 2004 in Beijing to participate in a ministerial meeting of the Shanghai Cooperation Organization (SCO), Russian foreign minister Ivanov and his Chinese counterpart confirmed their common approach to the settlement on the Korean peninsula and agreed to coordinate efforts in the interests of the political settlement. Valery Agarkov and Andrei Krylov, "Russia: FM Ivanov Ends Visit to China," *ITAR-TASS,* January 16, 2004.

of the fifth round of the six-party talks, Russian Deputy Foreign Minister Alexeyev and his Chinese counterpart Wu Dawei met to discuss key issues.[45]

Russia and China share almost identical goals and approaches with regard to North Korea's nuclear issue and have been closely coordinating their acts in order to bring the six-party talks to a successful conclusion. Russian Ambassador to China Sergei Razov stated that Russia and China maintained close coordination while drafting the UN Resolution 1718.[46] Russian Security Council Secretary Igor Ivanov also said that Russia and China were maintaining close contacts for a political settlement of the North Korean nuclear problem: "We have maintained and are maintaining permanent close contact with our Chinese partners, and our common goal is the nuclear-free status of the Korean Peninsula. Our work has been and will continue to be coordinated both at the bilateral level and in the six-nation format."[47]

Fully aware of its limitations, Russia refrains from attempting to take a leading role in North Korean nuclear talks. Instead, it seeks to play a supporting role in the negotiation process. It is fully aware that the main issues should be resolved between the U.S. and North Korea and wants China to exert its influence to bring these two antagonists back to the negotiating table and mediate a compromise to implement North Korea's denuclearization.[48]

Russia has attempted to bring about North Korea's denuclearization through unilateral, diplomatic means (mediation and good offices).[49] Russian leaders and academics postulate that North Korea's frequent and intimate contacts with Russia and its exposure to the outside world will prevent its isolation and reduce its nuclear ambitions.[50] They further argue that renewed friendship and trust between Russia and North Korea will help North Korea regain self-confidence and a self-confident North Korea will be better able to engage the outside world in a constructive way. In fact, Russia has been engaged in frequent dialogue with the DPRK through numerous channels. Russian leaders often spoke up on behalf of the DPRK at international gatherings, offered good offices between the two Koreas and between North Korea and Japan/the U.S., and made constructive suggestions to the DPRK, the U.S. and others.

---

45 "Spokesman Says Russia, China Have Common Vision of North Korea's Nuclear Problem," *ITAR-TASS,* November 8, 2005.

46 "Russian Ambassador: Russia, China Keep Closest Cooperation in Korean Nuke Issue," *Xinhua,* October 25, 2006.

47 "Russia, China Maintain Close Contacts Toward Settling DPRK Nuclear Problem," *Interfax,* October 31, 2006.

48 "Russian Official Hopes China to Play Active Role in North Korean Nuclear Case," *ITAR-TASS,* January 11, 2006.

49 For Russia's Korean policy in recent years, see Georgi Toloraya, "President Putin's Korean Policy," *The Journal of East Asian Affairs*, vol. 17, no. 1 (spring/summer 2003), pp. 33-51; Alexander Vorontsov, "Russia and the Korean Peninsula: Contemporary Realities and Prospects," *Far Eastern Affairs* (Moscow), vol. 30, no. 3 (2002), pp. 44-57.

50 Aleksey Bogaturov, deputy director of the Russian Academy of Sciences institute for the Problems of International Security stated: "Moscow is trying to prevent North Korea's further isolation" ("Putin Expected at Kim Chong-il's; Moscow Hopes To Avoid North Korea's Further Isolation," *Nezavisimaya Gazeta,* October 21, 2005).

President Putin made positive remarks on North Korean leader Kim Jong-Il at international gatherings. Immediately after his Pyongyang trip in July 2000, Putin characterized Kim as "a totally modern man" and "well-informed."[51] At the Okinawa G-8 summit, Putin also conveyed Kim Jong-Il's message that North Korea was ready to give up its current missile program if it could launch one or two peaceful satellites a year from the territory of other states with their assistance.[52] After returning from his trip to Pyongyang in August 2005, Russian presidential envoy to the Far Eastern Federal District Konstantin Pulikovsky in a press conference conveyed Kim Jong-Il's message that North Korea was ready to give up nuclear programs and to return to the NPT if its security was guaranteed: "Treat us [North Korea] like any other state, don't look at us as if we were an axis of evil and don't threaten us, and we'll not need these nuclear weapons at all. I'll not need a single nuclear warhead, or nuclear missile, if threats are removed from us." He further quoted the North Korean leader as saying, "given the current complicated economic situation, North Korea needs to develop its nuclear power industry and would appreciate if other countries did not raise obstacles in this work."[53] In regard to the agreements reached at the fourth round of the six-party talks, Pulikovski represented North Korea's position by saying that the DPRK would "fulfill the conditions set by the six-party talks on the Korean peninsula's nuclear problem, if the other parties involved fulfill them too" and that North Korea's demand for a light water reactor should be implemented.[54]

Russia also provided good offices between the DPRK and Japan. Japan reportedly requested Russia's help in resuming dialogue with North Korea,[55] and then Russian Foreign Minister Igor Ivanov spoke in July 2002 following his talks with Kim Jong-Il in Pyongyang, "Pyongyang is ready for a constructive dialogue with the U.S. and Japan without any preliminary conditions."[56] Moscow stressed that the success of the summit meeting between Kim Jong-Il and Japanese Prime Minister Koizumi Junichiro in Pyongyang in September 2002 was largely a result of Russia's mediation

51 *Itar-Tass*, July 20, 2000.

52 *Itar-Tass*, July 22, 2000, in FBIS, DR/SOV (2000-0722). In response to South Korean media reports that Kim Jong-Il proposal to cancel his missile program in exchange for commercial launches of Pyongyang's satellites by other countries was a joke, Russian Deputy Foreign Minister Alexander Losyukov stated: "our new contacts with the North Korean side left no impression that it was a joke." Kim Jong-Il made the proposal during the July meeting in Pyongyang with President Putin. *Itar-Tass*, September 11, 2000, in FBIS, DR/SOV (2000-0911)

53 "Russian Envoy: North Korea Leader Prepared To Give Up Nuclear Weapons Program," *INTERFAX*, August 18, 2005.

54 "Putin's Envoy Upholds North Korean Case Over US Reactor Pledge," *ITAR-TASS*, November 4, 2005.

55 "Japan Reportedly Asked Russia to 'Assist' in 'Resuming Dialogue' With DPRK," *Joongang Ilbo*, July 19, 2002, in FBIS, EAS/DR (2002-0719).

56 "Russian Foreign Minister Says N Korea Ready for Dialogue with US, Japan," *Itar-Tass*, July 29, 2002, in FBIS, DR/SOV (2002-0729).

policy, and Koizumi expressed his appreciation to Putin for his role in establishing Japan-North Korea dialogue.[57]

One day after North Korea's nuclear blast, Moscow expressed its willingness to join in multinational efforts toward a peaceful resolution of the crisis.[58] Commenting on North Korea's nuclear problem in December 2006, Russian Foreign Minister Sergei Lavrov called for compromise, goodwill, and patience: "But one can achieve the necessary compromises even on this situation, provided all the participants in the negotiating process have the goodwill and patience." He observed that Russia in cooperation with its partners managed to secure the unanimous adoption of UN Security Council resolutions 1695 and 1718, "which aim to bring the situation in the Korean peninsula back to normal through peaceful means on the basis of ensuring its nuclear-free status." He further stated that Russia would do its best to have the September 19 joint statement implemented: "Agreement to resume the six-party talks in Beijing, in the course of which the Russian side will do all it can to move forward towards the implementation of the joint statement of 19 September 2005 by finding solutions acceptable to all the participants, is also in line with these decisions."[59]

Concluding Remarks

With the convening of the six-party talks in 2003, Russia acquired the first opportunity to make its imprint on North Korea's nuclear issue. During the five rounds of nuclear negotiations spanning three and a half years (2003-2007) Russia has played a constructive role for negotiated resolution of North Korea's second nuclear crisis. Russian leaders made honest and sincere efforts to defuse tensions by offering good offices and serving as a messenger between North Korea and the US. Russia facilitated dialogue and offered suggestions for compromise. Russia at the same time exerted a moderating influence on both the U.S. and North Korea, constantly urging them to return to the negotiating table whenever the talks stalled and patiently reminding them to take into consideration the other side's concerns and needs during negotiation. North Korea's denuclearization and peaceful resolution of North Korea's nuclear crisis coincide with Russia's interests, and that's why Russia has been making genuine efforts to reduce tensions on the Korean peninsula and to bring an end to the nuclear crisis.

Russia's influence in the six-party talks has been indirect and marginal. The U.S. and the DPRK were the principal actors in this showdown and the DPRK as a host and mediator played a key role in inducing agreements especially during the fourth round of talks which resulted in the joint statement of principles. Russia's impact on the talks was limited to the format (six-party multinational talks) and the procedural matters (working-groups), and its impact on substance of the negotiated outcomes

---

57 "Russia Stresses Role in Mediating Between Japan, North Korea," *Itar-Tass*, October 11, 2002, in FBIS, DR/SOV (2002-1011).

58 "Russia ready to help in peaceful resolution of North Korean crisis," *RIA-Novosti,* October 10, 2006.

59 "Russia will strive for Korea talks compromise - foreign minister," *Interfax,* December 18, 2006.

was negligible. Russia exerted negative power throughout the talks. The outcome of negotiations depended largely on the U.S. and the DPRK and the diplomatic skills of China to bridge the gap between them. Russia along with China and South Korea set a certain boundary and stated that it would not accept any action or agreement beyond it. Russia carried out military exercises alone or jointly with China to show to the US and the DPRK that it had the military capabilities and the will to use them to protect its interests.

In the mid-1990s, Russian leaders realized that the U.S. was expanding its influence over the entire Korean Peninsula at the expense of Russia's legitimate security interests and that Russia was no longer considered a major player in Korean affairs because it lost influence over North Korea. Russian leaders calculated that Russia could regain its lost influence on the Korean peninsula by restoring friendly ties with Pyongyang while maintaining cooperative ties with Seoul. They also calculated that normalizing relations with Pyongyang would be beneficial to Russia's own security and economy. Has Russia now regained its lost influence over North Korea? The answer is a qualified "no." North Korea's primary concern is improving relations with the U.S. Its primary concern is to secure regime survival and improve the economic situation through direct negotiations with the US. Russia has limited influence over North Korea and its nuclear policies.

Chapter 8

U.S.-North Korean Negotiating Behavior and the Six-Party Talks

Scott Snyder

The experience of U.S.-DPRK negotiations during the first North Korean nuclear crisis has framed American and North Korean strategies and tactics as they have approached the second crisis. The Bush administration decided early in the second North Korean nuclear crisis to pursue a multilateral negotiating format (ultimately involving six parties—the two Koreas and four great power neighbors) based on its own perception of lessons learned from the Clinton administration's experience. These lessons included the desirability of negotiating with North Korea in a multilateral setting versus holding bilateral talks and the need to effectively counter DPRK negotiating tactics as a way of maintaining control of and denying the North Koreans opportunities to manipulate the negotiating agenda. For its part, the DPRK leadership has relied on a range of familiar approaches in the initial rounds of talks, including calculated efforts to utilize brinkmanship and crisis escalation in an effort to catalyze direct bilateral negotiations with the United States. Both sides attempted to shape the environment to their own ends and to isolate the other party as a vehicle for achieving their own strategic objectives through the mechanism of negotiations.

North Korea's nuclear test on October 9, 2006 appears to have marked a turning point that has affected the tactics of both the United States and the DPRK as the six-party talks have remained the primary vehicle for negotiating on this issue. Until North Korea's nuclear test, neither side had truly committed itself to resolving the nuclear issue through an active negotiation process. North Korea's nuclear test achieved Pyongyang's tactical goal of drawing the United States into direct talks, but against a strategic context that initially appeared to be relatively unfavorable to North Korea's longer-term strategic aims. The U.S. change in tactics appeared to concede many symbolic victories to North Korea as part of efforts to maintain international solidarity against North Korea's test and force Pyongyang to make a strategic change in course. This chapter will examine possible lessons learned from the first North Korean nuclear crisis, will review the decision to establish the six-party framework, and will analyze the tactics and strategies of the U.S. and DPRK, respectively, as well as the other parties to the talks.

The ABC's of the Bush Administration's Initial Approach to North Korea

The Bush administration came into office in January of 2001 convinced that it needed to take a very different approach to North Korea from that employed by the Clinton administration. Some of President Bush's closest advisors held strong views about the Clinton administration's approach to North Korea and about the Geneva Agreed Framework, a bilateral agreement negotiated between the United States and North Korea involving the construction of two light water reactors in North Korea in return for a North Korean freeze on its nuclear development and eventual dismantlement of North Korea's nuclear facilities. These views were based on perceived errors in the strategies and tactics of the Clinton administration and revealed a healthy skepticism about the process and outcome of any negotiations with North Korea. This policy approach to North Korea has been characterized as the Bush administration's "ABC" (Anything But Clinton) policy toward North Korea.[1]

The views of senior Bush administration officials on policy toward North Korea were shaped by frustration with and implemented in opposition to the policies of the Clinton administration in response to the first North Korean nuclear crisis of the mid-1990s. The visit of South Korean President Kim Dae-Jung in March of 2001 served to expose differences in the Bush administration's policy toward North Korea from both the views of South Korean allies and the previous approaches of the Clinton administration, as well as to expose deep fissures within the Bush administration over policy toward North Korea.[2] A subsequent policy review conducted following Kim Dae-Jung's visit and released in June of 2001 advocated a "comprehensive approach" for dealing with North Korea including "improved implementation of the Agreed Framework, a verifiable end to the DPRK's missile production and export programs, and a less threatening conventional military posture, but was not accompanied by any momentum or action plan to accomplish those goals.[3] The policy review masked policy differences, but served as a clear indication of the paralysis in Bush administration policy toward North Korea that resulted from deeply held internal disagreements over how to deal with the North.[4]

The events of September 11, 2001, served only to underscore differences between the United States and North Korea. These differences most dramatically came to be characterized by Vice President Cheney's statement that "we don't negotiate

---

1 Ivo Daalder and Bruce Lindsey, National Prospect, Peter Beck, "The Bush Administration's Failed North Korea Policy," JPRI Critique, Vol. XI, No. 3 (June 2004), accessed at http://www.jpri.org/publications/critiques/critique_XI_3.html.

2 Wendy Sherman, "Talking to the North Koreans," *New York Times*, March 7, 2001, p. 19; Michael R. Gordon, "How Politics Sank Accord on Missiles with North Korea," *New York Times*, March 6, 2001, p. 1.

3 James Kelly, "U.S. Policy in East Asia and the Pacific: Challenges and Priorities," Testimony before the Subcommittee on East Asia and the Pacific, House Committee on International Relations, Washington, DC, June 12, 2001. Available online at <http://www.state.gov/p/eap/rls/rm/2001/3677.htm> (accessed on February 2, 2007).

4 See Jane Perlez, "U.S. Will Restart Wide Negotiations with North Korea," *New York Times*, June 7, 2001, p. 1.

with evil, we defeat it."[5] Although North Korea publicly reaffirmed anti-terrorism commitments in response to those events, it became even more difficult to imagine a renewed U.S.-DPRK diplomatic dialogue at high levels. President Bush's pledge that "we will not allow the world's most dangerous regimes to threaten us with the world's most dangerous weapons" had further implications for U.S. policy toward North Korea as one of only a handful of potential enemy proliferators.[6] The initial presuppositions underlying the Bush administration's approach to North Korea were defined in opposition to the Clinton approach, North Korea remained a low priority, and the North Korean leadership became a convenient object of dramatic rhetoric through President Bush's personal comments about Kim Jong-Il and his characterization of North Korea as part of an "axis of evil" in his January 2002 State of the Union speech. The Bush administration's fundamental approach, thus, was one that dismissed the prospect of serious diplomatic interaction with Pyongyang and preferred to neglect the North Korean issue in favor of other priorities.

Despite the tougher rhetoric of the Bush administration, the North Korean side appeared to have high hopes that the Clinton administration's policy of rapprochement with North Korea would continue. Working-level officials of the U.S. State Department predicted to North Korean counterparts that eventually the Bush administration would pragmatically adjust and maintain a constructive dialogue with the North. Despite clear North Korean frustration with the rhetoric of the Bush administration, North Korean officials continued to focus on the need for an improved relationship with the United States as a primary policy objective and sought a senior-level dialogue with the Bush administration, anticipating as late as May of 2002 that the visit of a Bush administration special envoy may provide opportunities to stabilize U.S.-DPRK relations. Despite the Bush administration's harsh public rhetoric, Pyongyang's leadership held on to nostalgia for the Clinton days and to hopes that the Bush administration might eventually establish a high-level dialogue with the North, building on the Clinton administration's prior policies.[7]

The first high-level dialogue with North Korea during the Bush administration occurred in the context of Assistant Secretary of State for East Asia Jim Kelly's visit to Pyongyang as a special envoy in October of 2002, a trip that had been delayed from June of 2002 as a result of a military stand-off between the two Koreas in the West Sea. In the interim, the intelligence community had come to the conclusion that North Korea had indeed developed a covert uranium enrichment program for the purpose of developing nuclear weapons and Japanese Prime Minister Koizumi visited Pyongyang and gained the release of a small number of Japanese citizens who had been abducted from Japanese soil by North Korean agents in the 1970s and 1980s.

The Bush administration had indicated that it was prepared to pursue a "bold approach" in dealing with Pyongyang, but also raised troubling questions about the

5 Warren P. Strobel, "Vice President's Objections Blocked Planned North Korean Nuclear Talks," *Knight Ridder Washington Bureau*, December 20, 2003.

6 Remarks by President George W. Bush to Students of the Virginia Military Institute, April 17, 2002.

7 Author conversation with North Korean officials, May 2002.

need to resolve covert North Korean efforts to enrich uranium for use in nuclear weapons before such an approach could be adopted. Confronted with these allegations, the North Korean response was defiant and appeared to confirm the accuracy of the charges, although subsequent questions were raised over whether Kang Sok-Ju acknowledged the program or simply asserted that the DPRK was entitled to have a nuclear program in meetings with Kelly the day after U.S. concerns about North Korea's uranium program were presented. That contentious exchange marked the start of the second North Korean nuclear crisis, led to the unraveling of the Agreed Framework and North Korea's withdrawal from the Nuclear Non-Proliferation Treaty in late 2002 and January of 2003, and created a highly politicized and confrontational context in which the United States and North Korea have conducted diplomatic negotiations.[8]

Review of Six-party Talks, Rounds One-Five (2003-2005)

As the second North Korean nuclear crisis unfolded, three features of a new multilateral structure and process of negotiations emerged in contrast to the events of the first North Korean nuclear crisis, which was handled primarily through bilateral negotiations between Washington and Pyongyang. First, the United States was not eager to engage with North Korea bilaterally, preferring to create a multilateral context in which DPRK claims could be heard and responded to by multiple parties in an attempt to impose greater transparency, verifiability, and collective responsibility to the process. President Bush designed the issue as a "regional issue" early in the crisis, suggesting a quite different approach from one in which the United States might serve as chief guardian of a global non-proliferation regime.[9] Second, China emerged as a much more active participant, host, and mediator for what would become a six-party negotiation process, an approach that arguably provided China with greater leverage, responsibility, and influence to shape the outcome of the talks. Third, the six-party process created a dynamic in which the other parties besides the United States and North Korea were both witnesses and "jurists," weighing the respective claims of the United States and North Korea and creating a contest to isolate the other among the six parties as the impetus for movement in the negotiation process.

The first high-level diplomatic contact between the United States and North Korea following the Kelly visit occurred in trilateral talks hosted by the PRC in Beijing and held in April of 2003. At this preliminary meeting, the United States and North Korea held "talks about talks," that developed on the basis of Beijing's urgent efforts to bring the United States and North Korea to the negotiating table as a way of tamping

8 See David E. Sanger, "U.S. To Withdraw From Arms Accord With North Korea," *New York Times*, October 20, 2002, p. 1. For more details on the Kelly visit and a chronology of the crisis through 2005, see Scott Snyder, Ralph Cossa, and Brad Glosserman, "The Six-party Talks: Developing a Roadmap for Future Progress," online at <http://www.csis.org/media/csis/pubs/issuesinsights_v05n08.pdf>.

9 Federal News Service, Press Conference with President George W. Bush, East Room, The White House, March 6, 2003.

down a rising sense of crisis.[10] An immediate objective for Beijing was to ensure that a military confrontation did not develop between the United States and North Korea in the same way it was developing in the case of Iraq. Although no progress was made at this meeting, the dialogue itself provided a diplomatic vehicle for containing the North Korean nuclear crisis from escalating into a military confrontation. The North Korean side tried to raise the stakes through off-the-record comments that the North did indeed have nuclear weapons and that Pyongyang was prepared to further develop, produce, test, and/or export nuclear materials if the United States did not respond to North Korea's demands.[11] Following the April 2003 meeting, the United States sought to create a rhetorical consensus against North Korea's pursuit of nuclear weapons in summit meetings with leaders of Japan and South Korea and backed their inclusion multi-party negotiation process with North Korea. On the basis of effective trilateral coordination with allies, the Bush administration sought to reach out to China to host a multilateral negotiation process. In response, the North insisted that Russia also be included in the multilateral talks, and the six-party talks were born.

The first round of six-party talks held in August of 2003 served as an introductory session at which all parties expressed frustration with developments on the Korean peninsula, but little was accomplished beyond the parallel official statements presented at formal plenary sessions. At the second round in February of 2004, the United States pressed for the "comprehensive, verifiable, irreversible dismantlement" (CVID) of North Korea's nuclear program based on precedents set by the "Libyan model," in which Libya gave up its nuclear program in exchange for a more normalized relationship with the international community.[12] Not surprisingly, the North Korean side rejected the relevance of the Libyan model to the North Korean case and sought to exploit differences between the American insistence on non-proliferation and the concerns of South Korea and China, which were focused on regional stability with North Korea. South Korea and China perceived the American unwillingness to hold serious bilateral negotiations with North Korea as a primary obstacle to resolving the crisis. The North Koreans could leave Beijing confident that despite a rhetorical consensus against North Korea's nuclear program, the prospects for unified action to achieve that objective remained low and frustration with the Bush administration's hard-line approach remained high.

The third round of talks in June of 2004 marked the first time that the United States and North Korea submitted concrete proposals. The United States put forward a proposal that envisioned North Korea's "comprehensive dismantlement." Under the proposal, North Korea would commit to a freeze of its entire nuclear program as a first step toward dismantlement and within three months resume international inspections and monitoring to verify the freeze. In return, South Korea would

10 See The NewsHour with Jim Lehrer, April 16, 2003, "New Talks," Margaret Warner interview with Kurt Campbell and Henry Sokolski.

11 Guy Dinmore and Christopher Adams, "Bush Hits at Pyongyang's 'old blackmail game,'" *Financial Times*, April 26, 2003, p. 11.

12 Martin Parry, "U.S. Happy, But China says 'severe' differences remain as nuclear talks end," *Agence France Press*, February 28, 2004.

resume provision of energy supplies to North Korea and the United States and the DPRK would begin discussions on removing the DPRK from the U.S. government list of state sponsors of terror. In the second phase, the DPRK would receive multilateral security assurances that would be strengthened as the DPRK implements dismantlement and the United States and the DPRK would address other issues on the bilateral agenda.

The North Korean "freeze for compensation" proposal envisaged a limited freeze of North Korea's nuclear program with no preconditions in return for 2 million kilowatts of electricity and the lifting of U.S. economic sanctions on the DPRK. Inspectors would be drawn from the six countries participating in the six-party talks, and the scope of the freeze would only include plutonium (not uranium) generated following the North's withdrawal from the NPT. Although the tabling of the respective proposals might have signaled the beginning of a real negotiation process, it took over a year to return to the negotiating table.

In the interim, the North Koreans evaluated the result of the U.S. 2004 presidential election. Following the election, the DPRK foreign ministry released two dramatic unilateral declarations designed to enhance their strategic position and to recapture the attention of the United States. On February 10, 2005, North Korea declared that it was a nuclear state and on March 31, the North insisted that any return to negotiations would be on the basis of discussions of "mutual disarmament" with the United States. Although the Bush administration took a low-key approach to both of these statements, the groundwork was also laid for a return to six-party negotiations under the management of newly-appointed Assistant Secretary Christopher Hill.

The fourth round of talks finally convened following a direct meeting between Chris Hill and Kim Kye-gwan mediated by the Chinese just prior to a visit to Asia by Condoleeza Rice in July of 2005. The announcement of this meeting signaled that the United States adopted the tactical change of engaging North Korea directly in extensive bilateral negotiations over "principles," or parameters and objectives within which subsequent actions by each side might be undertaken. That intensive process, including numerous bilateral meetings among the parties, occurred from July 26 to August 7. Following a recess of five weeks, the talks reconvened on September 13, yielding a Joint Statement of principles on September 19, 2005.

The Joint Statement of September 19, 2005, addressed four separate areas where concrete actions might be taken to achieve the objective of the denuclearization of the Korean peninsula: denuclearization, normalization of U.S.-DPRK and Japan-DPRK relations, economic assistance, and the establishment of a permanent peace on the Korean peninsula. The Joint Statement also endorsed the principle of "words for words, actions for actions" as a basis for implementing the joint statement.[13] The expectation was that the Joint Statement would create a framework for progress in negotiating implementation of specific actions in fulfillment of the Joint Statement. A major disagreement in interpretation of the Joint Statement immediately overshadowed the agreement, as both the United States and the DPRK issued separate

---

13 Joint Statement of the Fourth Round of the Six-Party Talks, Beijing, September 19, 2005. See <http://www.state.gov/r/pa/prs/ps/2005/53490.htm> (accessed on February 2, 2007).

statements clarifying their respective positions on the sequencing of issues related to North Korea's right to have a peaceful nuclear program through the construction of light water reactors. The North made provision of LWRs a precondition of the implementation of the Joint Statement, while the U.S. statement indicated that a North Korean nuclear energy production program would only be considered after the North had fulfilled all of its denuclearization pledges.[14]

At the start of the fifth round of talks in November of 2005, the North Korean side accused the United States of violating the spirit of the joint statement by issuing "financial sanctions" against the DPRK. The "financial sanctions" were actually an advisory issued prior to the joint statement on September 15, 2005, naming the Macao-based Banco Delta Asia (BDA) as having facilitated money-laundering and counterfeiting activities through DPRK-affiliated bank accounts. This announcement caused a run on BDA and triggered a decision by the Macao Monetary Authorities to freeze the bank accounts in question and put BDA under its own supervisory authority. This issue became a major stumbling block and pretext for suspending renewal of the six-party talks. The November 2005 round of six-party talks was the last round of multilateral talks prior to North Korea's missile and nuclear tests on July 5 and October 9, 2006, respectively.

Lessons from the Geneva Agreed Framework and Assumptions Underlying the Bush Administration Approach to North Korea

Based on individual conversations with officials of the Bush administration and the broader course of action the Bush administration has taken in handling the second North Korean nuclear crisis, it is possible to determine some assumptions that Bush administration officials held as they dealt with North Korea. These assumptions might also be seen as parameters the Bush administration has employed in managing policy toward the North.

The first lesson the Bush administration appears to have drawn from the Clinton administration experience is that it is a bad idea to negotiate bilaterally with North Korea, and that only a regional solution to North Korean nuclear crisis is likely to be viable. The Clinton administration's bilateral approach to negotiating the Geneva Agreed Framework had been widely criticized for cutting out South Korea and Japan, parties whose security interests in North Korea's denuclearization were directly affected. The implementation of the Agreed Framework required Japanese and South Korean participation in any event, adding to the perception that a U.S.-led bilateral approach under the Clinton administration had played into North Korea's hands. In addition, the experience of KEDO negotiations with North Korea appeared to constrain some of the more dramatic and destructive aspects of North Korea's negotiating behavior since a multilateral negotiating approach limited the ability of Pyongyang to try to play differences among other parties off against each other.

---

14 See Jack Pritchard, "Six-party Talks Update: False Start or a Case for Optimism," Brookings Institution, The Changing Korean Peninsula and the Future of East Asia, December 1, 2005, see <http://www.brook.edu/fp/cnaps/events/20051201presentation.pdf> (accessed on January 29, 2007).

Moreover, high levels of mistrust between the Bush administration and the North Koreans made the prospect of unmediated bilateral negotiations highly unlikely to succeed. President Bush characterized the crisis as a "regional issue" at an early stage. The Bush administration's decision to promote the creation of the six-party talks as a multilateral channel by which to deal with North Korea appears to reflect these lessons from the first North Korean nuclear crisis.

A second conclusion the Bush administration appears to have drawn, which is particularly striking in the aftermath of 9/11, was that North Korean leadership/ Kim Jong-Il is rational and that Kim would not risk regime survival. There was relatively little concern that North Korea would take unconventional actions or reach out to terrorist groups as political allies since such steps might result in retaliatory attacks by the United States against North Korea similar to those against the Taliban following 9/11. The administration assessed that North Korea's capacity to directly affect U.S. security interests is limited and that deterrence, especially the threat of massive retaliation to North Korean aggression that would result in the end of North Korea, is sufficient to limit North Korea's strategic options.

The administration further determined that time was on the side of the United States. Secretary of State Colin Powell gave the impression in the early stages of the crisis that the United States would not be spurred to action by North Korea's reloading of its 5 megawatt reactor. Powell stated that "You can't eat plutonium. You can't eat enriched uranium. And as long as you pursue those technologies, those who can help you grow the things that you can eat, and develop an economy that will assist your people, can't help you."[15] Given North Korea's economic weakness, the administration appears further to have determined that North Korea has no other option but to try to negotiate a solution to the nuclear issue. As a result, the administration was willing to take its time waiting for North Korea to make a "strategic decision," and focused primarily on measures to enhance North Korea's political and economic isolation, not military threats, as the primary vehicles for attempting to influence North Korea to make a strategic decision.

A fourth assumption of the Bush administration was that "regime transformation" is a part of the solution to the nuclear crisis.[16] This formulation was useful because it masked a fundamental divide within the administration over whether the objective of the administration in managing the North Korean nuclear crisis should be "regime change" or a "change in the regime." In either case, the Bush administration continued to insist that North Korea must make a strategic choice, with the implication being that in the absence of a strategic decision to forego nuclear weapons, the leadership in North Korea would ultimately be unable to survive.

Fifth, the Bush administration seemed intent on countering well-known North Korean tactics, but in the absence of a realistic overall strategy for solving the crisis. The Bush administration we quite effective in countering many North Korean tactics including North Korean attempts to utilize brinkmanship and crisis escalation as means by which to set the agenda and time line for a negotiation process or to

---

15 Secretary Colin L. Powell Press Conference (as released by the State Department), Tokyo, Japan, February 23, 2003.

16 "U.S. Seeks NK Regime Transformation," *Korea Times*, December 9, 2004.

catalyze a U.S. response on terms favorable to the North Korean side. Instead, the Bush administration managed North Korean provocations by attempting to use them as means by which to reinforce North Korea's self-isolation. The assumption underlying this approach was that North Korea's pursuit of nuclear weapons would have the effect of further isolating North Korea from its neighbors. By attempting to remove the tactical gains that North Korea accrued from such behavior, the administration sought to induce a fundamental change in North Korean negotiating behavior. The administration's use of North Korean tactics as a means by which to enhance North Korea's isolation was incomplete, however, to the extent that others in the multilateral negotiating process continued to view American hard-line stances as the fundamental cause of the crisis. A further North Korean tactic that the U.S. side resisted was the North Korean effort to insist that others take measures to "save the face" of the North Korean side. Instead, it sometimes appeared that the Bush administration utilized excessive rhetoric to ensure that North Korean leaders lost face and did not gain legitimacy as a result of being "recognized" by the United States as a negotiating partner.

Finally, the administration may have also hoped that North Korea would be contained from further escalation by six-party Talks (especially China's influence to cap North Korean escalation steps). The six-party framework might be useful as a mechanism for crisis management, in addition to being (or until such time as circumstances permit it to be) a vehicle for multi-party negotiations. In the absence of a North Korean strategic decision, the Bush administration appears to have judged that further efforts to negotiate the abandonment of North Korea's nuclear program would be fruitless. But the existence of a multilateral framework would serve to manage the problem until conditions are more propitious for serious negotiation. This approach served to enlist the leverage of others, especially that of China, in the service of an effort to prevent North Korea from undertaking further escalatory measures in an attempt to use crisis to set the agenda and pace of negotiations.

The negotiation of the Joint Statement further enhanced the viability of the six-party process as a vehicle for crisis management since it committed the parties to pursue the shared objective of "the verifiable denuclearization of the Korean peninsula in a peaceful manner." However, the vagueness of the Joint Statement left room for disagreement on the process and priorities that would be emphasized in the course of fulfilling this objective. It is important to stress that this objective has two components: "denuclearization" and in a "peaceful manner." For many of the parties, the latter is as important as the former. The failure of six-party diplomacy risked the removal of a fundamental constraint on the use of military means to resolve peninsular issues, a course of action that would directly threaten Chinese and South Korean interests in regional stability and was also not in the U.S. interest.

North Korean Tactics and Strategy during the Second Nuclear Crisis

The task of discerning North Korean tactics and strategy remains extraordinarily difficult given the relative lack of opportunity to hear North Korean negotiators explain their internal negotiating strategies or to expound on their own thinking as they approach the North Korean nuclear issue. In the absence of credible opportunities to conduct interviews or receive direct explanations for the reasoning behind North Korean behavior, we can only interpret North Korean behavior based on the accumulated record of North Korea's past actions and any new explanations that the North Koreans themselves provide through public press statements. However, it is useful to examine North Korea's official statements and actions in the context of prior experience in dealing with the North as one way of understanding North Korean preferences and objectives in negotiations involving the United States.

During the first North Korean nuclear crisis, common tactics in the North's approach to negotiations have included the use of brinkmanship tactics and threats, crisis escalation tactics, and creating situations where other parties feel obliged to take actions to "save face" of North Korean counterparts. In addition, North Korean negotiators have consistently sought direct bilateral negotiations with the United States, both for prestige and legitimation purposes and as part of a strategy for maximizing gains and dividing parties with similar interests. Statements by the DPRK foreign ministry spokesman have been used to forward the negotiation process both through crisis escalation tactics and through active signaling of both North Korean concessions and essential preconditions for progress at the negotiating table.[17] Many of these same tactics have been a part of the North Korean response to the second nuclear crisis, especially the North's use of crisis escalation, threats, and blackmail tactics in an attempt to catalyze dialogue with the United States as well as efforts to both signal concessions, preconditions, and responses to U.S. positions at the negotiating table. The DPRK's use of public statements by the spokesman of the DPRK foreign ministry has been an important tool both for attempting to raise the stakes in the crisis and to outline the types of solutions that the DPRK would find acceptable throughout the crisis.

The initial DPRK foreign ministry statement concerning the visit by Assistant Secretary Kelly revealed North Korea's disappointment with his "high-handed and arrogant attitude" that made clear that the Bush administration is "pursuing not a policy of dialogue but a hardline policy of hostility to bring [North Korea] to its knees by force."[18] Although this statement did not reveal the core of the dispute that would spark a new crisis, it did reveal DPRK disappointment with the American approach to dialogue with the North since it precluded prospects for an improved U.S.-DPRK relationship that the North Koreans had desired to pursue since the end of the Clinton administration.

The first formal North Korean statement following public revelations from Washington of North Korea's apparent admission of a covert nuclear weapons program was issued in an October 25, 2002, statement, just prior to a meeting between President Bush and PRC President Jiang Zemin. It stated that the DPRK is "ready to seek a negotiated settlement" of the North Korean nuclear issue on condition that (1) the United States recognized North Korea's sovereignty, that (2) the United States assures North Korea of non-aggression through a treaty with the United States

17 See Scott Snyder, *Negotiating on the Edge: North Korean Negotiating Behavior* (U.S. Institute of Peace Press, 1999).

18 KCNA Statement, October 8, 2002.

and (3) that the United States does not hinder the economic development of North Korea.[19]

At this stage, the DPRK focused on opening a channel for direct negotiations under conditions favorable to the DPRK's own objectives, including assurances of non-aggression and an improved relationship with the United States. But the United States rejected these conditions as a basis for returning to negotiations and suspended provision of heavy fuel oil to North Korea though the Korean Peninsula Energy Development Organization (KEDO). For North Korea, the end of tangible direct benefits through KEDO set the stage for a series of escalatory actions by the DPRK, all announced through the foreign ministry spokesman from December of 2002 and through February of 2003. First, the DPRK foreign ministry announced that to cope with the electricity problem that the DPRK has "decided to lift the nuclear freeze."[20] North Korea kicked out IAEA monitors at Yongbyon, announced its withdrawal from the NPT, announced that it would reprocess fuel rods canned and safely stored under the Agreed Framework, and re-loaded its 5 megawatt nuclear reactor. All of these actions were signaled in advance and subsequently carried out as the DPRK sought to escalate the crisis as a means by which to set the stage for renewed negotiations in a context favorable to the North. But North Korea's combination of threats and crisis escalation maneuvers did not succeed in drawing the United States back into bilateral negotiations. Instead, the United States laid the ground work for multilateral talks in early 2003 by working with Japanese and South Korean allies and reaching out to China.

The North signaled a willingness to concede on the format of the talks in Beijing in April of 2003, but utilized additional bluffs and threats both at the Beijing meetings and in the aftermath of those meetings. North Korean assertions that they had nuclear weapons and were willing to transfer nuclear materials were tactics designed to escalate the crisis announcing that the DPRK is "compelled to unavoidably decide to equip itself with the necessary deterrent force" and "translated it into action," most probably by reprocessing the fuel rods that had been stored following the Agreed Framework. Following the first round of six-party talks in August of 2003, the DPRK expressed disappointment with the talks and reiterated its intention to "strengthen its nuclear deterrent" in an August 30 statement.

Following a statement by President Bush at the APEC summit in Thailand that the United States would be prepared to offer written security assurances, the DPRK's foreign ministry spokesman indicated a willingness to consider a U.S. "written non-aggression guarantee" if it is part of a "package settlement proposal" based on the "principle of simultaneous actions." In a November 16, 2003, statement, the DPRK foreign ministry spokesman further indicated that it would "abandon in practice" its nuclear program at the phase where the U.S. "hostile policy" is "fundamentally dropped" and its "threat" to the DPRK is "removed in practice," then further indicated the DPRK willingness to "at least agree on a first-stage action measure along with a word-for-word pledge at the next six-party talks," but insisted that such measures, including economic and political components, are needed in exchange for a "freeze"

19 KCNA foreign ministry statement, October 25, 2002.

20 KCNA foreign ministry statement, December 12, 2002.

of North Korean nuclear activities. He also states that North Korea would not freeze "nuclear activities for free, without any payment." These statements came to form the core of the "reward for freeze" proposal that North Korea eventually tabled at the third round of six-party talks held in June of 2004. (In addition, some of the language of statements by the DPRK foreign ministry spokesman has found its way into the Joint Statement issued at the end of the fourth round of six-party talks on September 19, 2005.)

A statement from the DPRK foreign ministry spokesman following the fourth round of six-party talks provides a relatively straightforward North Korean assessment of the U.S. and North Korean proposals and their relative strengths and weaknesses. It also gives a clear picture of factors the North has taken into consideration as it formulates its approach to the six-party process. The DPRK foreign ministry spokesman characterizes the DPRK proposal as one which requires a "big political decision for making a breakthrough" while advocating a "freeze" as the first step toward dismantlement and characterizes the size of the "reward" as the factor that would determine the ultimate timing of dismantlement. Regarding the American "landmark proposal," the DPRK foreign ministry spokesman characterized it as "complicated and unclear and unfair one as it lacked any U.S. commitment to implement the principle of "words for words" and "action for action." The spokesman illustrates the competition between the United States and North Korea for support among the other parties by noting that "the U.S. 'proposal' could not convince the participants in the talks ..."[21]

The two most dramatic uses of unilateral statements by the DPRK foreign ministry spokesman occurred in statements issued on February 10 and March 31, 2005. In the first statement, the DPRK announced the "suspension" of its participation in the six-party talks, and the second statement called for "mutual disarmament" talks with the United States. Both of these statements pursued several simultaneous North Korean objectives. First, the statement announcing the DPRK's suspension of its participation in six-party talks cited Bush administration rhetoric as a signal of U.S. continued interest in pursuing a policy of "regime change" in North Korea. Second, the statement utilizes the threat that the DPRK would "take a measure to bolster its nuclear weapons arsenal" as a self-protection measure.[22] By announcing the suspension of its participation in the six-party talks and the ongoing development of its "nuclear weapons arsenal," the North Korean side attempted to escalate a sense of crisis among the six-party talks participants and created a condition for return to the talks that the United States had to curb hostile rhetoric against the DPRK. The spokesman also attempted to create a justification for the DPRK to take another step in pursuit of its own nuclear program.

Three weeks later, the DPRK Foreign Ministry issued a lengthy "Memorandum" on March 3, 2005, putting forward its case for suspending talks and laying the basis for abandoning its pledge to adhere to a "missile test moratorium." The memorandum presents a set of conditions for the DPRK's return to the six-party

21 DPRK Foreign Ministry Spokesman on Six-party Talks, KCNA, June 28, 2004.

22 North Korea to "Suspend" Participation in Nuclear Talks, KCNA, February 10, 2005. As reported by BBC Monitoring International Reports from Nexis news service.

talks while attempting to further justify its decision to suspend participation in the talks and pursue "self-defensive" development nuclear and missile programs based on its perception of public manifestations of an American "hostile policy," including President Bush's call for an "end to tyranny" following Secretary of State Rice's characterization of the DPRK as an "outpost of tyranny." The DPRK calls upon the United States to "rebuild the groundwork of the six-party talks and create conditions and atmosphere for their resumption as quickly as possible," suggesting that the DPRK's "suspension" of its participation is a tactic designed to pressure and threaten the United States to quickly restore conditions for dialogue.[23] In a further attempt to shape the atmosphere in ways favorable to North Korea, the DPRK foreign ministry spokesman calls for "comprehensive denuclearization" in a March 31, 2005, statement.[24] Each of these statements attempted to catalyze a return to negotiations by escalating a sense of crisis, to create a situation in which the other party would have to save North Korea's face in the context of a perceived sleight or indignity, and to raise the bar for a return to the negotiation table in ways that would help to fulfill North Korean negotiating objectives and shape the agenda for future talks to its political advantage while enhancing DPRK prospects for extracting "rewards" from the talks.

In the ensuing months, Secretary of State Condoleeza Rice acknowledged that that the DPRK is a sovereign state and the United States reiterated that it did not have hostile intent against the DPRK, and Assistant Secretary of State Christopher Hill held a secret bilateral with his counterpart Kim Kye-gwan in Beijing, arranged by the PRC, the result of which was an announcement by the DPRK that it would return to a fourth round of six-party talks in August of 2005. This meeting, characterized by extensive bilateral negotiations between the United States and DPRK in the context of the six-party talks, resulted in a Joint Statement that contained many of the elements that the DPRK had sought, including an affirmation of the principle of simultaneity through "word for word, and action for action." The Joint Statement of principles also envisaged the denuclearization of the Korean peninsula, a normalization of diplomatic relations with the DPRK by the United States and Japan, economic benefits for North Korea, and a permanent peace settlement. These principles did not negate a "reward for freeze" component but did commit the United States to the principle of simultaneity as part of its approach to the North under the six-party framework. However, the announcement of the Joint Statement of Principles was accompanied by a familiar negotiation tactic that was used by both the United States and DPRK. Both sides agreed that the DPRK had a right to a nuclear program for peaceful purposes, but proceeded to define the wording of the Joint Statement in diametrically opposing ways that emphasized conditionality. The vagueness of the Joint Statement foreshadows the potential for extreme difficulty in carrying out the

---

23 "Memorandum of the DPRK Foreign Ministry," March 3, 2005. As reported by BBC Monitoring International Reports from Nexis news service.

24 "North Korea Calls for 'comprehensive denuclearization'—spokesman," KCNA, March 31, 2005. As reported by BBC Monitoring International Reports from Nexis news service.

task of determining specific actions to implement the principles embodied in the statement.

Following the negotiation of the Joint Statement, the emergence of the "financial sanctions" issue surrounding the U.S. Treasury's September 15, 2005, advisory regarding suspected counterfeiting and money laundering practices at BDA were taken by the North as a further signal of American ill-intent that violated the spirit of the Joint Statement. North Korean concerns about this issue expressed at the fifth round of six-party talks in November of 2005 led to a further stalemate in the six-party talks and opened the way for the North to pursue further brinkmanship and crisis escalation tactics, despite American efforts to ensure that such tactics resulted in the North's further self-isolation.

The North Korean Nuclear Test and the Resumption of Six-Party Talks

North Korea's July 5, 2006 missile tests and October 9, 2006 nuclear test appears to have marked a turning point that has affected the tactics of both the United States and DPRK. Thus far, the six-party talks have remained the primary vehicle for negotiating on this issue. But the North's actions also resulted in the passage of UN Security Council resolutions 1695 and 1718 condemning North Korea's actions and imposing limited sanctions on North Korea. Most dramatic was the extent to which China was willing to support these resolutions. Its strong statement of condemnation for the North's "brazen" act illustrated the extent of North Korea's isolation in the international community.

Despite North Korea's isolation, the nuclear test ironically worked as a catalyst for renewed six-party negotiations since none of the parties including China, South Korea, and the United States were willing to bear the costs that would accompany failed diplomacy. In fact, President Bush was notably restrained in his rhetoric toward the North Korean leadership for the first time drew a "red line" for North Korea by warning against the transfer of fissile material while insisting that diplomacy has not failed. In his statement responding to North Korea's test, President Bush warned that "The transfer of nuclear weapons or material by North Korea to states or non-state entities would be considered a grave threat to the United States, and we would hold North Korea fully accountable [for] the consequences of such action."[25] Days later, he drew a clear distinction with Iraq and repeatedly insisted that the proper response to the North Korean nuclear test was a renewed commitment to diplomacy.[26]

UN Security Council Resolution 1718 condemning the North Korean nuclear test was passed within only five days of the test, but the test also catalyzed diplomatic action by China to bring the situation under control. PRC President Hu Jintao condemned the North Korean tests as a "brazen" act, but simultaneously sent State Councilor Tang Jiaxuan as a special envoy to Washington, Moscow, and Pyongyang, where he became the first international visitor to meet with Kim Jong-Il following

25 Federal News Service, "Statement by President George W. Bush Re: North Korea's Nuclear Test," The White House, Washington, DC, October 9, 2006.

26 Federal News Service, "Press Conference with President George W. Bush," The Rose Garden, The White House, Washington, DC, October 11, 2006.

the nuclear test—in contrast to the aftermath of the missile test when a Chinese vice premier failed to see Kim Jong-Il. On his return to Beijing on October 20, he met with Secretary of State Condoleeza Rice who was on a swing through the region to rally support for implementation of UN Security Council resolution 1718. Less then two weeks following the meetings with Secretary Rice, Christopher Hill was back in Beijing at the invitation of the Chinese to meet with Kim Kye-gwan and to announce the resumption of six-party talks. After months of resisting bilateral negotiations unless the six-party process resumed, Chris Hill went to Beijing for discussions that led to the resumption of six-party talks. North Korea made a concession that facilitated the resumption of the six-party process by withdrawing the insistence that the "financial sanctions" issue had to be resolved before they would come back to the six-party talks, instead accepting that the issue would be addressed through a bilateral dialogue that would occur parallel to the six-party process.

The North Korean nuclear test was a strategic failure since Pyongyang came back to the talks more isolated than ever before, but it was a tactical success since it catalyzed renewed diplomacy and led the United States to reinitiate bilateral talks before the resumption of the six-party process. Chris Hill went back to Beijing for bilateral talks with Kim Kye-gwan on November 28-29, where he is reported to have presented an even more flexible and specific U.S. position on how to implement the pledges embodied in the 2005 Joint Statement. For the first time, the United States had embraced the principle of simultaneity in practical terms and was willing to consider a phased approach to North Korea's denuclearization that included a freeze as a first step and would be accompanied by political and economic rewards. Many elements of this approach appeared to be very similar to the North Korean proposal put forward at the third round of talks in June of 2004. Six-party talks resumed in Beijing in December of 2006 parallel with bilateral U.S.-DPRK talks on financial issues, but made limited progress. In mid-January, Kim Kye-gwan and Chris Hill met in Berlin for further discussions and the DPRK foreign ministry announced that a deal had been struck. Six-party talks resumed in mid-February of 2006 to bring the deal to a conclusion.

As of this writing, it appears that the United States utilized effective counter-tactics to blunt the most irritating aspects of North Korea's negotiating style, but has been unable to design an effective strategy to achieve North Korea's denuclearization. In the end the United States had no answer to North Korean crisis escalation tactics, which proved impossible to contain since North Korea inevitably would ratchet up the sense of crisis to higher levels than the United States would be willing to bear. As a result, these North Korean tactics were vindicated against the backdrop of the nuclear test, but at a cost. North Korea found itself more isolated than ever before.

Nor has the United States found an effective way to overcome the three factors that allow weak states to level the playing field in negotiations with much more powerful adversaries: alternatives, commitment, and control. To the extent that the North Koreans continue to show that they have alternatives to negotiation through unilateral pursuit of their own nuclear program, demonstrate commitment to their objectives, and maintain control over the nuclear materials and skills that are in dispute, it will prove exceedingly frustrating and difficult for the United States to neutralize the program.[27] Only through a convergence of common strategic purpose among all of North Korea's neighbors that outweighs the considerable mistrust among major regional powers will it be possible for an environment to be created in which the denuclearization of the Korean peninsula becomes a reality backed by action and not just a rhetorical slogan.

While American counter-tactics were initially successful, they also raised the bar of expectations for what was politically possible by demonizing North Korea. Yet in the absence of a military option, such demonization was extraordinarily unproductive. The Bush administration could not sweep the North Korean nuclear challenge under the rug nor did the administration have military or other leverage sufficient to bring the North Korean nuclear challenge to resolution. The effect was to raise the bar of expectations for what would be acceptable in dealing with North Korea far out of line with what would be possible in the context of a regional negotiation. One result is that the outcome will inevitably be disappointing, since the United States will inevitably be perceived as having given up far more than expected in order to gain a much lower threshold in terms of North Korean performance.

The North Korean side adhered to a more practical outcome for negotiations with its "reward for freeze" proposal than the American insistence on unconditional commitments to North Korea's denuclearization through CVID before being willing to provide North Korea with "rewards." As a practical matter, the North Korean step-by-step approach appeared to be the only way out, leaving all parties with the risk that the North Koreans might prefer to stop at an indefinite freeze as a way of preserving its nuclear status as well as its leverage in future negotiations.

It remained to be seen whether the rhetorical commitments to denuclearization of the Korean peninsula, which had grown increasingly stronger in light of the dangers posed by North Korea's nuclear test, would gradually become sufficiently ironclad that all parties in the region would adhere to their responsibilities and leave the North with no choice but to step back from nuclear brinkmanship. Now that North Korea has tested a nuclear device (although there are questions about whether the test succeeded or failed), North Korea is a de facto nuclear state. Is this a situation that North Korea's neighbors are willing to live with, or is a freeze on North Korea's capability to produce more weapons sufficient, given the lessons of the failed Agreed Framework? Only time will tell.

---

27 See Snyder, p. 69 and Mark Habeeb, *Power and Tactics in International Negotiation: How Weak Nations Negotiate with Strong Nations* (Baltimore, MD, Johns Hopkins University Press, 1988).

Chapter 9

Korean Peninsula Energy Development Organization (KEDO) A Bridge Too Far?

C. Kenneth Quinones

The Korean Peninsula Energy Development Organization (KEDO), initially was conceived as a joke, eventually became a reality but ultimately died a victim of intensifying mistrust between the United States and Japan, on the one hand, and North Korea. Established in March 1996, KEDO was the last of the confidence building measures initiated under the 1994 U.S.-North Korea Agreed Framework. After a slow and uncertain start, KEDO by 1999 became the core of an international effort to implement the terms of the Agreed Framework. Its primary purpose was to arrange for and oversee the construction of two light water nuclear reactors (LWR) at Shinpo on the northeast coast of North Korea. KEDO also had primary responsibility for annually supplying 500,000 metric tons of heavy fuel oil (HFO) to North Korea until the nuclear reactors had begun operation. Despite intense international effort and cooperation, as well as impressive progress toward its primary objective over one decade, KEDO died a quiet death in 2006. [1]

KEDO's experience merits close scrutiny today because the problem it was conceived to address persists. Some assume that North Korea and its persistent quest for a nuclear arsenal are the source of Northeast Asia's potential instability and the possible eruption of a second Korean War. It is a view firmly held in the incumbent Bush Administration and widely subscribed to in South Korea, Japan and the United States. Nevertheless, this assumption is highly debatable.

The "Korea Problem" and KEDO's Fate

North Korea's hostility and nuclear ambitions, it can be argued, are symptoms and not the source of the "Korea problem." Since the late nineteenth century, Korea has posed a serious problem to its neighbors. Militarily impotent and economically undeveloped, its neighbors feared that European imperialist nations would occupy the peninsula and use it as a base to expand their empire in East Asia. China clung to Korea as a tributary but did little to help it while worrying that it might fall prey to

1 KEDO's official reports can be found online at <www.kedo.org>.

Japan's apparent imperialistic impulses. Japan eventually concluded that a European imperial power might occupy feeble Korea and use it as a base for invading Japan. Japan, after wars with China and Russia, concluded it could solve the "Korea problem" by occupying Korea.

Colonization and Division

Japan's colonization of Korea, however, only complicated the "Korea problem." Suppression of Korean nationalist movements split Korea's political leadership into factions that eventually aligned themselves with rival superpowers, the United States and the Soviet Union. The Japanese imperial army's use of Korea as a base for the invasion of China ultimately set the stage for Korea's unwitting division in 1945 into two rival regimes, one loyal to the United States and the other to the Soviet Union.

North Korean leader Kim Il Sung decided in 1950 that he would solve the "Korean problem" by forcefully unifying the peninsula. His failure and the subsequent superpower rivalry of the Cold War crystallized Korea's division. But Koreans' belief that they could achieve national unification persisted. Koreans in both halves of Korea resolutely believe that the source of the "Korean problem" is their nation's division. When the Soviet Union collapsed in 1991, the people of South Korea confidently assumed that North Korea's collapse and unification were inevitable. Overlooked was the determination of North Korea's leaders to perpetuate their rule of Korea's northern half. Also overlooked was their determination to preserve their national sovereignty in the face of the superior military might of the U.S.-South Korea alliance.[2]

The Cold War's Legacy

The Cold War's end had only redefined the "Korea problem." For Pyongyang, the priority shifted from unification to regime survival. North Korean generals faced a strategic choice. They could accept disarmament, and pursue reconciliation with South Korea, according to the terms Seoul and Washington set, or they could reinforce their nation's "deterrent capability."

Initially Pyongyang opted to negotiate with South Korea, a process initiated in 1990. Despite impressive progress, however, intense distrust and rivalry between the two Koreas persisted. North Korea's generals' recognized that the Soviet Union's collapse had ended its protective "nuclear umbrella." At the same time, the United States' decisive and quick victory in the first Gulf War using "smart" weapons rendered obsolete all the Soviet-era military equipment that North Korea relied upon to equip its million-plus man army.

2 For a discussion of the prelude to the U.S.-North Korea nuclear negotiations of 1993-94 that formulated the Agreed Framework, see C. Kenneth Quinones, "North Korea: From Containment to Engagement," in Dae-sook Suh and Chae-jin Lee, eds., *North Korea After Kim Il Sung* (London: Lynne Rienner Publishers, Inc., 1998).

The Agreed Framework

In 1993, North Korea's leadership opted to negotiate with the United States. It offered to trade its nuclear weapons development program for a series of simultaneous steps that would culminate in the normalization of bilateral U.S.-North Korea diplomatic and commercial relations. The effort yielded the Agreed Framework. Signed on October 21, 1994, this was the first U.S.-DPRK bilateral diplomatic agreement.[3]

It was formulated on the assumption that mutual distrust could gradually be erased by a process of simultaneous steps designed to build confidence and trust between the United States and North Korea. At the same time, the United States deemed it equally important that North and South Korea engage in and expand their dialogue to similarly replace their mutual distrust and rivalry with a process of reconciliation.

Central to both processes was implementation of several confidence building measures. These included:

- The United States supply of heavy fuel oil (HFO) to generate an amount of electricity to replace that North Korea would not be able to generate after it shut down its nuclear reactor at its Yongbyong Nuclear Research Center.

- The placement of North Korea's 8,000 nuclear spent fuel rods in long term storage under the International Atomic Energy Agency's (IAEA) monitoring.

- In exchange, North Korea would freeze all its nuclear related activities, remain a member of the Treaty on the Non-proliferation of Nuclear Weapons (NPT), and allow IAEA monitoring and inspections of its nuclear facilities.

- The establishment of diplomatic liaison offices in each nation's capital to facilitate diplomatic communication and negotiation vital to the agreement's implementation.

- The United States would phase out economic sanctions imposed on North Korea since the start of the Korean War in 1950. This would occur as North Korea demonstrated its cooperation in the locatation and recovery of the remains of some 8,100 American military personnel who died in North Korea during the Korean War, and proved the credibility of its claim that it no longer supported international terrorism.

- The United States promised to organize and head an international organization to facilitate construction of two light-water nuclear reactors (LWR) at Shinpo on North Korea's east coast.

- Once completed, North Korea pledged that it would allow the United States to ship the 8,000 spent nuclear fuel rods to a third nation and permit the dismantlement of its nuclear reactor.

3 Arms Control Association, "The U.S.-North Korea Agreed Framework at a Glance,: www.armscontrol.org/factsheets.

Even before the Agreed Framework had been concluded, however, South Korea's political leader President Kim Yong-sam, took issue with the Agreed Framework. He believed that North Korea's economic woes and estrangement from the international community would lead to its collapse. He favored a strategy of containment that would reinforce North Korea's economic problems and isolation in the belief that this would inevitably and quickly result it the North Korean regime's collapse. This stance obviously clashed with the Agreed Framework's basic assumptions.

Seoul verse Washington

Also, no sooner had the Agreed Framework been signed than the Republican Party won the mid-term Congressional election of November 1994. This gave the Republicans control of Congress and put the Democratic Clinton Administration on the political defensive. Republicans quickly aligned themselves with President Kim Yong-sam's goal of working to bring about the collapse of the Kim Jong-Il regime in Pyongyang. They shared the belief that the Agreed Framework and Clinton Administration's other confidence building measures with North Korea, including humanitarian assistance, prolonged Kim Jong-Il's rule.[4]

Within this politically charged environment, implementation of the Agreed Framework began. By 1998, the Agreed Framework was on the verge of collapse because of opposition to it in Seoul and Washington. Its life was prolonged when Kim Dae-Jung was inaugurated South Korea's new president in February 1998. Fundamental to his so-called "Sunshine Diplomacy" was the engagement of North Korea in a series of confidence building measures that would promote reconciliation. Consequently, his administration strongly supported the Agreed Framework.

The reverse was true in Washington and Pyongyang. By 1998, the Republican controlled Congress imposed increasing fiscal controls on the weakened Clinton Administration's ability to fund the programs vital to the Agreed Framework's successful implementation. At the same time, Pyongyang's critics of the agreement claimed that the United States had failed to fulfill its part of the agreement. They also argued that United States policy toward North Korea was increasingly hostile rather than moving toward normal relations. As evidence, they pointed to U.S. Congressional laws designed to restrain funding and implementation of the agreement.[5]

---

4 C. Kenneth Quinones, "South Korea's Approaches to North Korea: A Glacial Process," in Kyung-Ae Park and Dalchong Kim, *Korean Security Dynamics in Transition* (New York: Palgrave, 2002).

5 Richard Armitage, "The Armitage Report on North Korea: A Comprehensive Approach to North Korea," *Congressional Record* (March 4, 1999), pp. 341-343. U.S. Congress, "House-Senate Conference Report on HR 4328 (Omnibus Appropriations Bill), Sec. 582. Korean Peninsula Energy Development Organization," See *Congressional Record* (October 19, 1998).

Pyongyang Blunders

North Korea blundered on August 31, 1998 when it launched a multiple stage ballistic missile over Japan's main island. If successful, the missile could drop a warhead on the United States. The launch shattered confidence in the United States and Japan that the Agreed Framework could replace mistrust with normal diplomatic relations. But by 1998, the "Korea Problem" once again had become one of intense mutual mistrust. Pyongyang's refusal to end its missile program and growing suspicions about its cooperation with Pakistan regarding nuclear weapons technology fed this distrust in Washington and Tokyo.

The Clinton Administration redoubled its efforts in 1999 and 2000 to reassure the U.S. Congress and people that a process of engaging North Korea could convert it into a respectable member of the international community. North Korea's September 1999 moratorium on the testing of ballistic missiles helped, as did President Clinton's appointment of former Defense Secretary William Perry as his North Korea Policy Coordinator. President Kim Dae-Jung's "sunshine diplomacy" prolonged the building of two light water reactors in North Korea.

The first North-South Korea summit of June 2000 revived hopes that the Agreed Framework could succeed. Rather than continuing to cling to the past, the two Koreas resolved that they would pursue a hesitant process of reconciliation. Shortly afterward, the Clinton Administration took unprecedented steps toward reconciliation with North Korea. In October, 2000, North Korean leader Kim Jong-Il dispatched his senior most general to Washington for high level talks and President Clinton soon after sent Secretary of State Albright to Pyongyang. The visits, however, yielded no enduring progress toward reconciliation.

Mistrust Intensifies

The underlying reason remained mutual distrust. In Washington, some in the intelligence community believed that North Korea had initiated a clandestine program to develop highly enriched uranium. If true, this was contrary to pledges it had made as part of the 1994 Agreed Framework and the earlier 1992 Joint North-South Declaration on the Denuclearization of the Korean Peninsula. In Tokyo, North Korea's August 1998 ballistic missile test had frightened and angered the Japanese people. Further reinforcing their distrust of North Korea was Pyongyang's previous abduction of many Japanese citizens and continuing insistence that it had not done so. In Seoul, Kim Jong-Il's reluctance to fulfill his 2000 North-South Korea Summit pledge to visit South Korea convinced many in the south of his insincerity regarding reconciliation.

Soon after the U.S. Supreme Court had designated him the victor in the 2000 U.S. presidential election, George W. Bush made numerous comments and took a series of steps that appear to have convinced North Korea's generals that the United States had decided to undermine North Korea either using economic sanctions or military means. For reasons unlikely to be fully known for decades, North Korea's leaders in January 2003 determined that the Agreed Framework had failed and they opted to pursue the development of nuclear weapons.

Despite three years of intense diplomatic effort between 2003 and 2006 by China and South Korea, the pursuit of a peaceful diplomatic end to North Korea's nuclear program has thus far failed to achieve any concrete results. On the contrary, the Agreed Framework has been discarded and North Korea has tested a nuclear explosive device.

Japan and the United States reject the notion that the problem is mistrust. Instead, they insist that Pyongyang's reluctance to fulfill its promises to the international community and quest for weapons of mass destruction is at the heart of the "Korea problem." In the event that a new bilateral agreement is eventually formulated between Washington, DC and Pyongyang, no matter its terms, ultimate successful implementation will hinge on the extent to which these two adversaries achieve mutual trust. Without it, no agreement can succeed.

It is against this back drop that the saga of persistent mistrust that the KEDO must be understood. The entire story of the Agreed Framework's demise is too complex to relate in depth here. Thus we will concentrate on a single bilateral endeavor created by the Agreed Framework: KEDO.

The Agreed Framework's Faltering Start

KEDO was the last of the confidence building measures to be implemented as part of the 1994 Agreed Framework. Simply put, implementation of the agreement proved far more challenging than its negotiation. It took only sixteen months to formulate the accord, but a decade of intense effort failed to achieve its successful implementation.

Promises to Congress

Robert Gallucci, the U.S. chief negotiator of the Agreed Framework, and his staff concentrated initially on dealing with Congress and seeking funds to implement the Agreed Framework. Candidly speaking, their efforts yielded disappointing results, a consequence more because of Republican opposition than a lack of earnestness. The Republican Party's success in the November 1994 mid-term elections proved a formidable obstacle to their efforts. But Gallucci and his staff nevertheless proved unable to rally broad Congressional support. A major reason was Gallucci's promise to the Congress in November 1994 that the Agreed Framework's implementation would be paid for by Washington's allies South Korea and Japan as part of their "burden sharing" of the cost of maintaining U.S. military forces in Northeast Asia. Subsequently, Congress held the Executive Branch to its word and only relatively small amounts of funding were authorized for the accord's implementation. This problem haunted the Agreed Framework until its demise in 2006.[6]

---

6 Much of the information about the early implementation of the Agreed Framework, KEDO's establishment, etc. is based on the author's personal experience first as the North Korea officer in the Office of Korea Affairs, Bureau of East Asian and Pacific Affairs, then as a member of the U.S. Spent Fuel Team working in North Korea and the North Korea political analyst for the State Department's Bureau of Intelligence and Research, East Asia Division.

Gallucci's pledge also displeased South Korea and Japan. At the time, East Asian Affairs Deputy Assistant Secretary Thomas Hubbard closely advised Bob Gallucci on how to deal with the South Korean and Japanese governments. Hubbard and Gallucci shared the dated view that the two U.S. allies should concede that their funding of the Agreed Framework was consistent with their "burden sharing" responsibilities. While the two allies generally shared this view, Washington's rather presumptuous attitude in this regard excited criticism in Seoul and Tokyo. Politicians in both capital felt their governments had been slighted when Washington assumed that they would cover 90 per cent of cost of building two nuclear reactors for North Korea. This legacy also severely handicapped the agreement's implementation until 1998 when South Korean President Kim Dae-Jung assumed primary responsibility for funding the construction project.

Burden Sharing in Washington

The Agreed Framework's implementation in Washington from the beginning was hampered by the lack of staff and funding. While Ambassador Gallucci and his staff initially focused on dealing with Congress, the burden of implementation vis-à-vis North Korea fell to the understaffed Office of Korea Affairs at the State Department and other agencies: the Arms Control and Disarmament Agency (ACDA), and the Departments of Defense (DoD) and Energy (DoE). These latter agencies supplemented the East Asia Bureau's meager personnel and funding resources. On the other hand, these agencies knew little or nothing about how to deal with North Korea, which could be said about the entire U.S. government at the time.

From September 1993 to October, 1994, the "Korea Desk" had only three effective officers to deal with both North and South Korea. The office director was of little help since he had never dealt with either Korea. The heaviest burden fell on the Deputy Director Lynn Turk, who spoke very good Korean and had extensive experience in South Korea and elsewhere in East Asia, and the North Korea affairs officer Kenneth Quinones, the only U.S. diplomat who had visited North Korea until that time. Lynn Turk was compelled to concentrate on South Korea because the South Korea affairs officers were both in a state of duress. One's daughter had died from a long illness and the other was experiencing severe emotional problems. Unfortunately, the Office of Personnel failed for unknown reasons to provided qualified replacements.

Help did not arrive at the Korea Desk until after the Agreed Framework's signing. But Quinones moved in September 1994 to the Bureau of Intelligence and Research so that he could continue his liaison duties with the North Korean government. Three officers replaced him on the "Korea desk," but none had any experience dealing with North Korea. At the same time, the office director initiated preparations to retire from the Foreign Service.

Meanwhile, Gallucci was given the rank of ambassador and became the chief coordinator of North Korea policy. His focus eventually shifted early in 1995 to the establishment of KEDO while Hubbard worked with the South and North Korean governments to resolve their differences regarding the type of reactor to be built in North Korea.

KEDO'S Cousins

During the 1994-95 winter months, implementation of the accords confidence building measures fell to Turk and Quinones. Turk took charge of the U.S.-DPRK liaison office negotiations and related matters. Assisting him were two Foreign Service officers newly assigned to the Korea Desk. Quinones concentrated on providing the Secretary of State with political analysis about North Korea's intentions while advising ACDA officials on how to deal with the North Koreans and to make arrangements for the first U.S. visit to the Yongbyon Nuclear Research Center.

Prior to KEDO's formal establishment in March 1995, Turk initiated negotiations with North Korea regarding the establishment of liaison offices by each side in the other nation's capital. It was hoped that the opening of these offices would facilitate direct diplomatic dialogue, ease mutual suspicions and promote cooperation vital for implementation of the agreement. The first negotiations convened in Washington in early December 1994.

The talks went well, but a blunder by a U.S. Army helicopter pilot undermined the progress and excited mutual mistrust. When an American Army helicopter pilot became disoriented and flew his aircraft into North Korean airspace, the North Korean People's Army (KPA) promptly blew it out of the sky. The pilot was killed and his co-pilot taken prisoner. The ensuring crisis was resolved when Assistant Secretary Hubbard traveled to North Korea and signed an apology regarding the incident. Thereafter, however, progress toward the opening of liaison offices never resumed.

The other preliminary confidence building measure also quickly encountered formidable obstacles, but not in North Korea. Many months before KEDO was born, a team of representatives from the Arms Control and Disarmament Agency (ACDA), State Department officers (including the author), representatives of the Department of Energy and some nuclear technicians met their North Korean counterparts in Berlin, Germany in September 1994. Their objective was to determine how best to initiate implementation of three key elements of the Agreed Framework:

1. How would the U.S. compensate North Korea for shutting down its nuclear reactor at the Yongbyon Nuclear Research Center located about one hundred kilometers due north of Pyongyang?
2. How would the U.S. place the 8,000 spent nuclear fuel rods at Yongbyon into long term safe storage; and
3. What kind of nuclear reactor would the U.S. provide to North Korea?

North Korea insisted on receiving compensation for the electricity that it would lose by shutting down its Yongbyon nuclear reactor. The U.S. offered to build two conventional thermo power plants fueled by coal to generate an amount of electricity. North Korea's chief negotiator, Kim Jong-u, then chairman of the DPRK's Committee for the Promotion of External Trade, adamantly rejected this. Ultimately, the U.S. conceded to the North Korean demand that the U.S. annually supply 500,000 metric tons of HFO until the reactors to be built as part of the Agreed Framework went on line and began producing electricity.

Heavy Fuel Oil

The U.S. also agreed to deliver the HFO on a scheduled basis to the port of Najin-Sonbong. The Soviet Union had years before built a thermo power plant there that burned HFO to generate electricity. The plant was close to North Korea's industrial corridor that ran along the nation's northeast coast.

The Berlin Talks, however, failed to achieve agreement on the type of nuclear reactor to be built in North Korea. South Korea insisted that its version of a U.S. based reactor be provided to North Korea since the U.S. expected Seoul to pay most of the reactor's construction costs. North Korea adamantly rejected the offer and insisted upon a reactor that was completely of U.S. design and origin. The disagreement did not obstruct the signing of the Agreed Framework, but it required a year of intense negotiations between KEDO and Pyongyang before the DPRK agreed to accept the South Korean design. Again, Washington's taxing of Seoul to pay for the Agreed Framework further complicated and delayed the Agreed Framework's implementation.[7]

Nevertheless, the Berlin Talks of September 1994 set the stage for the successful signing of the Agreed Framework in October 1994. The U.S. promptly called on the DPRK to engage in negotiations in Pyongyang about how to put the nuclear spent fuel into long term storage. The talks convened in early November 1994 and made impressive progress. The U.S. delegation was also able to visit Yongbyon, and take measurements and photographs to facilitate the design of equipment that would be needed to carry out what came to be known as the "spent fuel canning project."

Back in Washington, however, the lack of funds continued to hamper the accord's implementation. The U.S. had promised the first delivery of HFO to the DPRK in December, 1994. But there were no funds to purchase the HFO until the Defense Department agreed to contribute $5,000,000 toward the purchase. A South Korean supplier made the delivery and a potential crisis was averted. Nevertheless, a serious problem was highlighted – there was no money available to purchase and pay for the shipment of HFO to North Korea.

Spent Fuel Project

At the same time, the Spent Fuel Project encountered the same reality. While North Korea was cooperating fully, Washington was unable to respond in kind because there was to money to fund its promises. As for KEDO, it remained more a dream than a reality.

The Spent Fuel Project nevertheless went forward. Relying on ACDA and DoE funding, the U.S. negotiating team returned to Pyongyang in January 1995

---

7 The author participated in this September 1994 talks and maintained a personal journal.

to finalize the details of agreements and procedures that would govern the project. After the successful talks, a small team consisting of two American scientists, a State Department officer and an interpreter went to Yongbyon and worked there until early February. This work determined the essential needs for the spent fuel project. At the same time, DPRK officials in the Atomic Energy Agency, Korean People's Army (which controlled access to Yongbyon), and the Foreign Ministry further confirmed their government's willingness to facilitate implementation of the Agreed Framework.

But when the American team returned to Washington in February, 1995, they learned that further progress would have to wait funding. In March, 1995, the U.S. State Department representative on the Spent Fuel Project, Kenneth Quinones was dispatched to Congress to brief key House staff on the need for the funding and progress made thus far. The House Foreign Relations Committee Majority Staffer Peter Brookes summarily dismissed Quinones' briefing and insisted that Congress had already been told that no U.S. funding would be needed to implement the Agreed Framework.

It was not until June 1995, primarily because of repeated effort by the Departments of Defense and Energy that sufficient funds were found to initiate the Spent Fuel Project. Initially the U.S. government had promised in November 1994 that the project would commence in December. It was not until July 1995 that the project's essential equipment began to arrive at the Yongbyon Nuclear Research Center. The project was finally completed and all 8,000 spent fuel rods placed in safe storage in October 1997. At no time during the project did the DPRK government obstruct the project. The primary impediment from the beginning had been the lack of U.S. funds for the project which ultimately cost less than $30 million.

KEDO's Early Life

Whereas members of the Spent Fuel Team had to endure the bitter cold of Pyongyang and Yongbyon in January, KEDO's founders worked in comfortable offices in Washington and enjoyed wining and dining with their South Korean and Japanese colleagues. Then, as KEDO took form, the harsh early realities of initiating the Agreed Framework's implementation in North Korea faded from memory and the focus shifted to the much more comfortable endeavors in Washington and KEDO's New York office.

The KEDO concept was first suggested on a warm, humid July day in Geneva during the second round of U.S.-DPRK nuclear in 1993. DPRK chief negotiator and First Vice Minister of Foreign Affairs Kang Sok-ju surprised his U.S. counterparts. After a long and unproductive morning session, Kang returned from lunch to proclaim that his government was prepared to make a "magnanimous" offer. Pyongyang would remain a member of the Treaty on the Non-proliferation of Nuclear Weapons (NPT) and cooperate with the IAEA if the United States would provide the DPRK two light water nuclear reactors.

No one on the U.S. team thought such a proposal could ever win support in Washington, D.C. The proposal was virtually identical with the Soviet Union's 1985 promise to the DPRK. In exchange for North Korea's signing the NPT, Moscow would build two nuclear reactors in the DPRK. Pyongyang did sign the treaty and Moscow did conduct a site survey at Shinpo on North Korea's east coast, but the deal collapsed along with the Soviet Union.

When Kang proved adamant in his proposal, U.S. chief negotiator Gallucci promised to do his best to promote the idea in Washington, D.C. He cautioned, however, that winning Washington's support for the idea would be extremely difficult. Prior to the U.S. negotiating team's return to Washington, late night and early morning team chat sessions entertained suggestion on how the idea might be presented back home. Gradually, a consensus centered on the name "KEDO," Korean Energy Development Organization. As for funding, early on it was suggested that South Korea and Japan could be the primary funding partners. After all, they would be the primary beneficiaries of any agreement that halted North Korea's nuclear weapons program. Their contribution would be labeled part of their "burden sharing" with the United States' efforts to ensure their national security.

The "KEDO" idea, however, did not win substantial support until the summer of 1994. In June, the United States and North Korea had come very close to colliding militarily over the nuclear issue. The Clinton Administration, under pressure at home to establish its foreign policy credentials with a major success, was increasingly eager to close a deal with Pyongyang. Seoul added to the pressure. President Kim Yong-sam was anxious to reduce tensions on the Korean peninsula and to preserve his nation's prosperity by avoiding a clash with North Korea. Japan's government shared similar sentiments.

The "KEDO" concept won support in July and August 1994 when Gallucci and Hubbard visited Tokyo and Seoul. Tokyo sanctioned the concept but hesitated to commit to funding the project. ROK President Kim, however, offered to fund most of the nuclear reactor project, i.e. KEDO, because he saw it as a way to take control of the post-agreement process and use it to compel North Korea's leadership to deal with him.

Surprisingly, KEDO by September 1994 had won the diplomatic support of Washington and Tokyo, and substantial financial support from Seoul. That left the need to win Pyongyang's cooperation. As always, Pyongyang proved a hard sell. It was not until December 1995, one year after it had accepted the KEDO concept and nine months after KEDO formal birth in March that KEDO finally won Pyongyang's cooperation.

Pyongyang's mistrust of Washington and Seoul initially was the foremost obstacle to KEDO's birth. From the start, North Korea insisted that the United States must have primary responsibility for the LWRs' construction. Also, it demanded that Washington serve as the LWR project's primary contact point. When Pyongyang learned that Washington expected Seoul to be the major source of funds, it objected. Matters became increasingly complex when Pyongyang learned that Washington supported Seoul's insistence that the reactors to be built for North Korea would be partially of South Korean design. Diplomatic dueling over these issues consumed a year of haggling and were not resolved until December 1995 when North Korea finally signed the reactor supply agreement with KEDO.[8]

Uneven but Steady Progress

The spent fuel canning and HFO projects continued their uneven progress into 1996. Funding was finally authorized for the "canning" project. Several tons of custom designed equipment began arriving at Yongbyon in the summer of 1996. The first priority went to "vacuuming" highly radioactive sludge from the bottom of the spent fuel pond. This was necessary to ensure clarity of the water so that the spent fuel rods, each about one meter long, could be counted. Then huge steel "boxes" were positioned on the bottom of the pool. These were to hold canisters that would each contain 21 fuel rods. Work platforms were assembled and placed into the storage pond.

North Korean technicians had to be trained first in nuclear safety practices. They were taught the importance of wearing protective clothing. Then they learned how to operate the equipment that would first pick up a single fuel rod, insert it into a cleaning mechanism and then put the rod into a storage canister. Once filled, the canister was sealed and positioned in the large steel "boxes." When a box had been filled, specially designed seals locked the canisters into the storage boxes.

Kerosene fueled generators, water filtration machines, vacuum machines, canisters, work stations, video cameras, an entire laboratory, safety equipment and clothing, etc. were all fabricated in the United States. They were then flown by air cargo to Beijing, China where everything was transferred to cargo aircraft leased from the North Korean air force. Each flight cost $30,000, paid in cash in advance, for the flight to Pyongyang. Trucks were then loaded and driven the one hundred kilometers to Yongbyon.

The entire "canning" project required three years to complete. It began in January 1995 with analysis of the problem and finally concluded in the fall of 1997. Every detail of the process was negotiated and formalized in a set of "Record of Meetings." Work procedures, living arrangements, transportation and logistic, payment of expenses for personnel assistance, medical care, communications, etc. - every detail of the operation was negotiated in a series of sessions that began in November 1994 and ended in the summer of 1996. During the project's three years, six to eight American contract workers lived and worked at the Yongbyon Nuclear Research Center. A State or Energy Department volunteer often visited the site to resolve any disputes and to assess progress.

---

8 See "Agreement on the Establishment of the Korean Peninsula Energy Development Organization (March 9, 1995); "Joint U.S.-DORK Press Statement (Kuala Lumpur, June 13, 1995); and "Agreement on Supply of a Light-water Reactor Project to the DPRK Between KEDO and the Government of the DPRK (December 15, 1995). Also see Center for Nonproliferation Studies, Monterey Institute for International Studies, "Korean Peninsula Energy Development Organization." Monterey, CA: 2002.

The project was a dress rehearsal for KEDO's LWR project. North Korea's cooperation was impressive to say the least. Numerous obstacles, misunderstandings and projects had to be overcome, but for the most part North Korean authorities worked cooperatively with their American counterparts.

Despite the "canning" project's unprecedented success, KEDO officials largely ignored it, but the North Koreans did not. They used the agreements forged for the "canning" project as precedent for the protocols formulated with KEDO.

HFO's Sputtering Start

Despite KEDO's June 1995 pledge to make regular deliveries of HFO, KEDO continued to struggle to supply HFO between 1995 and 1997. The problem was the lack of funds to purchase the HFO and to pay the cost of delivering it. Initially loans secured by the Japanese government enabled KEDO to supply limited amounts of HFO to North Korea. Then the State Department dispatched Ambassador Paul Cleveland and his assistant on numerous funding raising trips to ask for contributions to KEDO's HFO fund. Several nations, including Russia, offered to help. Russia's offer was rejected, but offers from New Zealand, Australia, Canada, Chile, Argentina and eventually the European Union gained them membership in KEDO.[9]

North Korea repeatedly blamed its shortage of electricity on KEDO's uneven supply of HFO. Such claims, however, were groundless. North Korea's energy crisis was a consequence of its antiquated electricity generating and distribution facilities, and had nothing to do with KEDO's shortcomings. But blaming North Korea's problems on foreigners distanced Pyongyang's leadership from these problems and cast them as victims rather than the source of the problem.

Actually North Korea struggled to consume all the HFO it received. It insisted that the deliveries continue to be made to its northeast port of Najin-Sonbong. But the old power plants there simply could not burn the HFO. They broke down repeatedly and, because they were of Soviet-era design, repair was difficult to say the least. Eventually North Korea found it necessary to dig huge holes in the earth where the HFO was dumped once storage tanks had filled.

KEDO continued to struggle for survival until 1999. The lack of political support in Washington, D.C. and Seoul impeded sufficient funding for all of KEDO's projects. Contributions from Japan and other KEDO members prevented KEDO's collapse until President Kim Dae-jung assumed primary responsibility for the organization's funding.

---

9 See KEDO's *Annual Reports, 2001-2004.* In June 1995, KEDO and North Korea's Foreign Ministry agreed upon a schedule for HFO deliveries. KEDO proved unable to keep the schedule. The author was present when this agreement was formulated and monitored KEDO's HFO deliveries from June 1995 until his departure from the State Department in October 1997.

Humanitarian Assistance

KEDO also benefited indirectly from the international community's generosity in the form of food aid. North Korea's grain harvests began declining in the late 1980s. By 1994, the shortage of rice and corn had become a serious concern. Initially South Korea and Japan addressed North Korea's food shortage with large gifts of rice in the summer 0f 1995. But torrential rains in August destroyed much of North Korea's grain harvest and the nation experienced a period of sporadic famine and pervasive food shortages that lasted until 1999.[10]

The United States was one of many nations that provided humanitarian assistance to North Korea between 1995 and 2000. It quickly became the largest donor of food aid by 1996. American based humanitarian non-government organization's (NGOs) also pioneered a diverse variety of humanitarian projects in North Korea. These activities received U.S. government encouragement and, in July 1997, were formalized into the Private Voluntary Organization Consortium or PVOC. Six to eight representatives of different NGOs took up residence in Pyongyang to monitor North Korea's distribution of U.S. government funded food aid. The U.S. Agency of International Development paid the PVOC's operational costs. The program proved highly successful, albeit somewhat controversial, until the summer of 2000 when it's Congressional critics halted funding for it.

The North Korean government greatly appreciated the humanitarian aid provided by the United States, both the government and private organizations. The effort fostered widespread goodwill for Americans in North Korea. It may even have restrained Pyongyang's criticism of the United States' faltering implementation of the Agreed Framework.

Pyongyang Blunders—the 1996 Submarine Incident

Pyongyang proved unable or unwilling to restrain its mistrust of President Kim Yong-sam and his administration. For unknown reasons North Korea dispatched a commando-laden submarine to South Korea's east coast in September 1996. After the commandoes had disembarked on the coast, the submarine ran aground. Despite South Korea's impressive and extensive defense establishment, the submarine was discovered by a taxi driver. Subsequent discovery that North Korean commandoes had landed sparked outrage and near panic in South Korea. When all but one of the commandoes was found dead and the survivor apprehended, the immediate crisis subsided.

The incident had pervasive political consequences. Foremost, the incident rekindled suspicions about North Korea's trustworthiness and its ultimate intentions. It reinforced the view that North Korea was pursuing a duel track of appearing to transform itself into a respected member of the international community but at the

---

10 C. Kenneth Quinones, "The American NGO Experience in North Korea – A Preliminary Assessment, *Proceedings of the First World Congress on Korean Studies,* The Academy of Korean Studies (July 2002). Marl Manyin, "U.S. Assistance to North Korea: Fact Sheet," Washington, D.C.: Congressional Research Service, 2005.

same time continuing its orthodox unsavory conduct. President Kim, supported by the people of South Korea, lashed out at North Korea for its insincerity and duplicity. The United States and Japan promptly echoed South Korea's outrage and shut down all projects with North Korea. The activities of KEDO, the spent fuel canning project and the U.S. Army's search and recovery of Korean War remains were all suspended. Only an unprecedented North Korean public apology, negotiated by the United States, to South Korea resolved the matter. Distrust of North Korea, however, lingered, particularly among Washington and Seoul's critics of the Agreed Framework.

KEDO Goes to Work

KEDO did not actually begin work at the Kumho construction site until its ground breaking ceremony on August 19, 1997, more than three years after its founding. No sooner had it initiated the construction of the LWRs than the Asian Financial Crisis spread to South Korea. The crisis, sparked by the huge debts East Asia's rapidly growing economies had development over the previous decade, thoroughly disrupted government spending in Thailand, South Korea and other previously prosperous developing economies. Intervention by the International Monetary Fund (IMF) prevented a global financial crisis and stabilized the situation. The IMF also attached mandatory reforms to the loans it made to the adversely affected economies.

KEDO's work was further disrupted by the crisis. The South Korean government had given the Preliminary Work Contract (PWC) to the Korea Electric Power Company (KEPCO) to prepare the Kumho site in North Korea for the LWR project. Port facilities had to be built, workers dormitories, offices and cafeteria had to be constructed, electricity generating facilities built and roads paved. Despite the financial crisis, KEPCO in December 1997 began its work with the understanding that it would be duly compensated.

Sunshine Diplomacy

Kim Dae-Jung's victory in South Korea 1997 presidential election ushered in "sunshine diplomacy" and a new era in North-South Korea relations. President Kim's efforts undoubtedly prolonged the life of the Agreed Framework and KEDO.

By August 1997, North Korea's foreign ministry informed the State Department that Pyongyang was running out of patience with Washington's uneven implementation of the Agreed Framework. North Korean diplomats were telling their American counterparts that their leader Kim Jong-Il was listening more closely to the criticism and advice of his generals than his diplomats.

Pyongyang had long used the flow of HFO as an indication of Washington's earnestness in fulfilling its promises under the Agreed Framework. The unsteady flow of HFO had become increasingly unreliable, something Pyongyang's critics of the accord repeatedly cited as evidence of waning U.S. commitment. Pyongyang repeatedly asked for the U.S. president or secretary of state to make a public statement supportive of the Agreed Framework. But President Clinton was then preoccupied with his own personal and political crisis while the secretary of state concentrated on other pressing matters. At the same time, there was an increasingly vocal chorus of critics in Congress who wanted the United States to distance itself from the Agreed Framework. Pyongyang's August 31, 1998 launching of a Taepodong multi-stage ballistic missile over northern Honshu Island of Japan intensified criticism of the Agreed Framework in Washington, Tokyo and Seoul. By then it had become viewed as "appeasement" that prolonged the rule of Kim Jong-Il's despotic regime. President Kim had resolved KEDO's financial crisis, but thereafter the organization's survival increasingly hinged on politics in Washington.

President Kim Dae-Jung's "sunshine diplomacy" took root in this hostile atmosphere. Quite astutely, he first garnered international support for his strategy during visits to Beijing, Tokyo, Moscow and Washington. For the first time, South Korea had won support from all four superpowers for its strategy to deal with North Korea. While advocating resolute support for maintaining military deterrence aimed at North Korea, he sought to engage North Korea in a conciliatory manner. One of his more important steps in this regard was to pledge that his administration would fund 70 per cent of the estimated $4.6 billion cost of the KEDO LWR construction cost. Japan joined by promising $1 billion and the European Union offered Euro 75 million for KEDO's operational costs over a five year period. At long last, almost four years after its establishment, KEDO achieve a sound financial basis.

Slippery Slope to Demise

The Clinton Administration sought to align itself more closely with Kim Dae-Jung's sunshine diplomacy in 1999 and 2000. The effort was admirable but possibly too tardy to reverse the U.S. Congress' increasing skepticism about the Agreed Framework. Pyongyang meanwhile did little to salvage the situation.

North Korea's military exchanges with Pakistan and refusal to negotiate a deal to end its ballistic missile exports intensified suspicions in Washington regarding Pyongyang's ultimate intentions. Some in Washington concluded that Pyongyang was pretending to be nice to gain food and other material assistance while secretly pursuing nuclear weapons development. This suspicion leaked into the American press in August 1998 and excited allegations that North Korea was building a secret underground nuclear site. Eventual U.S. inspect of the site in May 1999 proved the suspicions were groundless.

Nevertheless, the suspicions persisted. Some in Congress accused North Korea of diverting HFO for use by its armed forces. KEDO's installation of meters to measure the flow of HFO into and out of North Korea's storage tanks demonstrated that such diversion was marginal at most. Similarly, North Korea was accused of diverting U.S. food aid to the North Korean army instead of the intended beneficiaries, North Korea's civilian population. Again, monitoring by American humanitarian works and an increasing number of international monitors proved that any diversion was relatively insignificant. North Korea was not entirely innocent in every case, but equally certain is the fact that most of the allegations against it were exaggerated.

The Republican controlled Congress, however, passed measures designed to limit the extent to which President Clinton could fund the Agreed Framework's implementation and supply food aid to North Korea. Rather than building mutual trust, the Agreed Framework's implementation because the source of growing mistrust.[11]

KEDO forged ahead. On December 15, 1999, KEDO finally signed its turnkey contract with KEPCO. At long last, actual construction of an LWR in North Korea was scheduled to commence on February 3, 2000. Unfortunately for KEDO and its project, a protocol negotiated earlier between KEDO and North Korea had projected completion of the first LWR in 2003. Originally, 2003 had been estimated as the completion date based upon initial assumptions that construction would commence in 1995 or 1996. But when that proved impossible, the original date was never adjusted. Since the turnkey contract was not signed until December 1999, at least three years later than originally anticipated, completion of first LWR by 2003 would be impossible.[12]

North Korea subsequently argued that the United States and/or KEDO should compensate it for the tardy LWR delivery by building North Korea two conventional thermo power plants. Pyongyang also demanded that KEDO provide the funds needed to modernize North Korea's dilapidated and outdated electrical grid system. These demands were repeatedly rejected, but they further fueled criticism in Washington and Seoul of the Agreed Framework.

The growing chorus of critics in Washington was reflected in the so-called "Armitage Report" of March 1999. Former Defense Assistant Secretary Armitage had convened a working group in Washington that consisted of selected Republican critics of the Agreed Framework and the Clinton Administration's approaches to North Korea. Included were prominent "neo-conservatives" like Ambassador Paul Wolfowitz, House International Relations Majority Staffer Peter Brookes, and future National Security Council East Asia Adviser Michael Green, among others. Their report claimed that, "Arguably, the Agreed Framework was a necessary but not sufficient response to the multiple security challenges posed by North Korea." It claimed that the Agreed Framework was based on faulty assumptions that enabled

11 See note five above. Also see U.S. House of Representative's Report 105-825, "Making Omnibus Consolidated and Emergency Supplemental Appropriations for Fiscal Year 1999, Sec. 582. Korean Peninsula Energy Development Organization. Online at < http://thomas.loc.gov>; and U.S. Institute of Peace, *Special Report: Mistrust and the Korean Peninsula – Dangers of Miscalculation.* online at <www.usip.org>; and Global Reporting Network, "Politics and the Agreed Framework: North Korean Deal on Thin Ice? Online at <www.bu.edu/globalbeat>.

12 <www.kedo.org>. In October 1994, just after the Agreed Framework was signed, the author and then Assistant Secretary Gallucci's deputy drafted the first schedule of anticipated mile stones in the agreement's implementation. We estimated, based upon construction starting no later than 1996, the first LWR would be completed by 2003. Unfortunately, KEDO staff negotiated the "Agreement on Supply of a LWR Project to the DPRK …," without adjusting the date for the LWR's completion. See the December 15, 1995 agreement, Article III, paragraph 1 which states, "KEDO shall develop a delivery schedule for the LWR project aimed at achieving a completion date of 2003." Beginning in 1998, KEDO inability to fulfill this promise became a major irritant in its relationship with North Korea.

North Korea to endure as a threat to the United States and its allies Japan and South Korea. Congressman Benjamin Gilman, chairman of the House International Relations Committee, embraced the "Armitage Report," as did numerous other influential members of Congress.[13]

Flickering Hope

Pyongyang belatedly attempted to revive trust with Washington toward the end of 1999. After numerous inconclusive rounds of talks aimed at ending North Korea's export of ballistic missiles, Pyongyang finally pledged to stop testing its ballistic missiles during talks with the United States in Berlin in September 1999. Looking back, the gesture now appears to have been too little too late to revive U.S. Congressional support for the Agreed Framework.

Of greater consequence was the first North-South Korean summit that convened in Pyongyang in June 2000. The meeting between South Korean President Kim Dae-Jung and North Korea's Supreme Commander Kim Jong-Il surprised and thrilled the international community. It suggested that sunshine diplomacy could work and that North Korea could be nudged toward becoming a respectable member of the international community.

Washington reacted by trying to play catch up to South Korea's reconciliation strategy. Pyongyang and Washington exchanged high level visits in October 2000. The meetings, their symbolism and rhetoric suggested that the Agreed Framework could still work. But two developments quickly deflated this optimism.

In November 2000, representatives of the United States and North Korea met in Kuala Lumpur to try once again to forge an agreement to end North Koreans ballistic missile sales and development. The meeting once again ended inconclusively. Pyongyang reportedly refused to specify which ballistic missiles it would stop exporting and developing.

A few weeks later, the Republican Party claimed victory in the 2000 U.S. presidential election. The new U.S. President during the campaign had repeatedly alleged that the Agreed Framework was an example of Clinton Administration "appeasement," termed North Korea a "failed state," and labeled its ruler a tyrant. Jokingly, but quite accurately, "Korea watchers labeled new Bush Administration's approach toward North Korea "ABC," or "anything but Clinton."[14]

President Bush's rhetoric toward North Korea and its leadership first eroded the waning trust previously forged by the Agreed Framework, and then intensified mutual hostility. In December 2001, President Bush identified North Korea as a potential target of his new "pre-emptive strike" strategy aimed at countering the proliferation of weapons of mass destruction by claiming that the United States had the right to attack any nation that it felt threatened its national security. Then in his January 2002 State of the Union speech, he identified North Korea as one of four

13 See notes five and eleven above.

14 C. Kenneth Quinones, "Dualism in the Bush Administration's North Korea Policy," *Asian Perspective,* vol. 27, no. 1 (2003), pp. 197-224.

nations be alleged belonged to an "axis of evil." Meanwhile suspicions that North Korea had secretly resumed its nuclear weapons development program, a violation of the Agreed Framework if accurate, convinced President Bush to issue an April 1, 2002 Presidential Determination in which he declined to certify that North Korea was in compliance with the Agreed Framework. This action suspended further U.S. funding for KEDO.[15]

KEDO's efforts peaked in 2002 and then rapidly declined. In April, KEDO's contractor KEPCO completed construction of dock facilities at Kumho. This allowed the transport by sea of heavy equipment and construction materials from South Korea directly to the LWR construction site. By August 2002, KEPCO began pouring concrete for the first LWR's foundation at Kumho.

Bush Administration distrust of Kim Jong-Il's government climaxed, however, in October 2002. A U.S. diplomatic delegation headed by State Department Assistant Secretary of State for East Asian and Pacific Affairs James Kelly traveled to Pyongyang in early October 2002. This was the Bush Administration's first endeavor to engage North Korea in substantive diplomatic dialogue. When Kelly asked his North Korean counterpart Kim Kye-gwan whether Pyongyang had initiated a clandestine program to produce highly enriched uranium (HEU), Kim initially denied this. (North Korea had previously pledged not to engage in such activities as part of the Agreed Framework which also mandated that North Korea observe the terms of the 1992 Joint North-South Declaration on the De-nuclearization of the Korean Peninsula.)

When the talks resumed the next day, North Korea's first Vice Minister of Foreign Affairs and chief negotiator at the Agreed Framework talks, Kang Sok-ju, reportedly reading from a prepared statement, admitted that his government had an HEU program. The outraged U.S. delegation abruptly walked out of the talks. Unfortunately no one on the U.S. side attempted to obtain a copy of Kang's statement. Later North Korea issued a formal public statement denying that Kang had made such an admission. But it was too late. The U.S. delegation had departed Pyongyang without seeking further clarification. Mistrust on both sides had finally boiled over into mutual hostility.[16]

Back in Washington, D.C., the Bush Administration kept the matter quiet until October 16. National Security Council adviser on nuclear proliferation affairs, Robert Joseph, "leaked" the story to selected journalists. He claimed that the Bush Administration had determined that North Korea had committed a "material breech" of the Agreed Framework. That afternoon Joseph briefed journalists from the *New*

15 Presidential Determination No. 2002-12 (April 1, 2002). online at <www.presidency.ucsb.edu>. Also see U.S. Congress, 1st Session, January 13, 2003, S.145 to prohibit certain assistance to North Korea or the Korean Peninsula Development Organization and other purposes. online at <http://thomas.loc.gov>.

16 Based on conversations with three of the delegation's members.

York Times, Chicago Tribune, U.S.A Today, Washington Post and Los Angeles Times.[17]

Thereafter, the Bush Administration successfully convinced KEDO's board of directors that resolute action was needed to convince North Korea to halt its clandestine nuclear activities and return to compliance with the Agreed Framework. On November 14, KEDO suspended its HFO shipments to North Korea. One week later, KEDO halted all its activities, including construction of the LWR at Kumho.

South Korea's financial support for KEDO continued until 2006, but KEDO never resumed its work in North Korea. At first, North Korea allowed KEDO to maintain a token staff at the construction site, but no work was conducted. Finally when KEDO ceased to exist in 2006, North Korea claimed ownership of the partially constructed LWR and its facilities.

Conclusion

KEDO was the child and victim of adversity. The organization was inseparable from the agreement that established it and the political will that gradually forged it into a reality. The KEDO concept—an international organization designed to replace a nuclear weapons program with a peaceful nuclear power program—was innovative to say the least. It was also a highly creative effort to bridge the half century gap between Korea's two belligerent halves. Looking back to 1994, when KEDO was first conceived, we must recall that North-South Korea relations consisted of intense animosity and mistrust. Dialogue had given way to hostility and collaboration to rivalry.

The Agreed Framework successfully bridged that gap between 1994 and 2000. It began by halting North Korea's nuclear weapons program, restoring international monitoring of its secret facilities and tempering hostility between Pyongyang and Washington. The framework's several confidence building measures, beginning with the spent fuel canning project in January 1995, demonstrated that Americans and North Koreans could work together and accomplish tasks of mutual benefit. The Defense Department's Joint Recovery Operations with the Korean People's Army added further evidence of this beginning in June 1996. Eventually KEDO contributed further evidence that the Agreed Framework could bridge the gap of mistrust between Washington and Pyongyang, and even between Seoul and Pyongyang.

Ultimately, the Agreed Framework, including KEDO, proved unable to bridge the gap. The problem was not the agreement or its confidence building measures such as KEDO. Rather, from the beginning, the accord's implementation was severely hampered by a lack of political commitment in Washington and Seoul. While Washington was reluctant to fund the project, Seoul was unwilling to support the project politically. This eroded Pyongyang's commitment to the agreement. Subsequent blunders by Pyongyang rekindled mistrust.

17 Based on conversations with one of the journalists involved.

Ultimately the Agreed Framework and KEDO collapsed because of a lack of trust between Washington, Seoul and Pyongyang. The accord and KEDO had successfully built trust between 1994 and 2000, but the politicians in each capital had contributed to the erosion of that trust.

Korea Peninsula Energy Development Organization (KEDO)

Chronology

September 15, 1994	U.S.-DPRK talks in Berlin regarding type of nuclear reactor and heavy fuel oil to be supplied to the DPRK.
October 21, 1994	U.S.-DPRK Agreed Framework signed.
December 5, 1994	First shipment of heavy fuel oil (HFO) arrives at Najin-sonbong, DPRK.
December 20, 1994	North Korea's Army shot down an U.S. Army helicopter over North Korea.
January 1995	Four member U.S. scientific team, including State Department Officer C. Kenneth Quinones, begins three week stay at Yongbyon Nuclear Research Center.
March 9, 1995	Korea Peninsula Development Organization established in New York City. (Name later changed to Korea Peninsula Energy Development Organization.)
March 19, 1995	First KEDO board meeting convenes in New York. Ambassador Stephen Bosworth named first director.
July 1995	U.S. Spent Fuel Team installed equipment and generators at Yongbyon Nuclear Research Center to initiate "canning" process.
September 1995	U.S. Spent Fuel team commences vacuuming sludge from spent fuel storage pond at Yongybyon.
	UN World Food Program (WFP) opens an office in Pyongyang and U.S. State Department dispatches humanitarian aid to the DPRK.
December 15, 1995	KEDO-DPRK Reactor Supply contract signed.
September 1996	North Korean submarine was discovered grounded on

	South Korea's east coast.
January 1997	U.S. Spent Fuel Team begins placement of spent fuel rods (canning) in long term safe storage under International Atomic Energy Agency (IAEA) monitoring.
	Asian Financial Crisis sweeps across Southeast and East Asia, severely disrupting all commercial and government activities in the region.
	KEDO-DPRK protocols governing site take-over, logistics and non-payment arrangements concluded.
July 1997	U.S. Agency for International Development authorizes Mercy Corps International to organize the Private Voluntary Organizations Consortium (PVOC) to monitor the distribution of U.S. government food aid in the DPRK. Representatives of six U.S. humanitarian non-governmental organizations (NGO) take up residency in Pyongyang.
August 19,1997	Four Party Talks between South and North Korea, China and the USA convene to address the Korean Peninsula's problems.
	KEDO conducts ground breaking ceremony at Kumho nuclear reactor site in North Korea.
October 1997	U.S. Spent Fuel Team completes "canning" at Yongbyon.
December 1997	KEDO and Korea Electric Power Company (KEPCO) sign Preliminary Work Contract (PWC) to commence work at Kumho reactor site.
February 25, 1998	Kim Dae-Jung inaugurated ROK president and initiates "sunshine diplomacy." President Kim does not continue the Four Party Talks.
August 31, 1998	DPRK's first Taepodong multi-stage missile test fails.
November 9, 1998	KEDO Board of Director agrees on a division of costs for the reactor construction project. Of the estimated \$4.6 billion cost, South Korea agrees to fund 70%, Japan pledges \$1 billion and the European Union

	pledges to provide Euro 75 million for operational costs over a five year period.
November 12, 1998	Former Defense Secretary William Perry appointed North Korea Policy Coordinator.
March 4, 1999	The so-called "Armitage Report" severely criticizes Clinton Administration policy toward North Korea.
April 25, 1999	First Trilateral Coordination and Oversight Group (TCOG) meeting between the U.S., ROK and Japan convenes.
Sept. 7-12, 1999	U.S.-DPRK ballistic missile talks produce DPRK Moratorium on the testing of ballistic missiles.
Sept. 15, 1999	William Perry delivers report to Congress.
December 15, 1999	KEDO-KEPCO signed reactor construction contract which took effect February 3, 2000.
June 15, 2000	ROK President Kim Dae-Jung meets DPRK Supreme Commander Kim Jong-Il in Pyongyang.
June 19, 2000	U.S. relaxes selected sanctions on the DPRK.
Oct. 9-12, 2000	DPRK Field Marshal Jo Myong-rok visits Washington, D.C.
Oct. 22-24, 2000	U.S. Secretary of State Albright visits Pyongyang.
January 20, 2001	President Bush inaugurated.
February 2001	Bush initiates review of North Korea policy.
March 2001	Presidents Bush and Kim Dae-Jung hold first summit.
December 2001	Bush Administration declares right to pre-emptive attack on any nation that threatens U.S. national security. The DPRK is designated a potential target.
January 21, 2002	Bush declares the DPRK a member of the "axis of evil."
April 1, 2002	Bush refuses to certify that North Korea is in compliance with the Agreed Framework, halting U.S. funding for KEDO.

August 7, 2002	KEDO begins pouring concrete for reactor foundation at Kumho.
Oct. 3-5, 2002	U.S. State Department Assistant Secretary Kelly heads a delegation to Pyongyang and is told that the DPRK has a highly enriched uranium (HEU) program in violation of both the 1992 Joint North-South Korea Declaration on the De-nuclearization of the Korean Peninsula. And the U.S.-DPRK Agreed Framework.
Oct. 16, 2002	Bush Administration declares that the DPRK is in "material" violation of the Agreed Framework.
Nov. 14, 2002	KEDO suspends heavy fuel oil (HFO) shipments to the DPRK.
Dec. 12, 2002	DPRK informs the IAEA it will restart it nuclear reactor.
Dec. 27, 2002	DPRK expels IAEA inspectors.
Jan. 10, 2003	DPRK withdraws from the Treaty on the Non-proliferation of Nuclear Weapons (NPT).
Aug. 27-29, 2003	Six Party Talks between China, Japan, the two Koreas, Russia and the U.S.A commence in Beijing to seek a peaceful diplomatic end toNorth Korea's nuclear program.
Nov. 21, 2003	KEDO Board of Directors suspends all activities effective for one year beginning December 1, 2003.
January 6, 2006	KEDO withdrew all its personnel from the Kumho construction site located thirty kilometers north of Sinpo, South Hamgyong Province on North Korea's east coast.
	Soon after KEDO ceased to exist as a functioning entity.

Chapter 10

The Six-Party Talks and Security Cooperation in Northeast Asia

Yong-Sup Han

In the twenty-first century, countries in Northeast Asia are pursuing a new regional security and economic order. During the first half of the twentieth century, they lived in an age ridden with wars and poverty, whereas, in the latter half, they began to speed up economic growth in an age of peace and prosperity. Nevertheless, the expectation that the twenty-first century will become an age of peaceful coexistence and co-prosperity in Northeast Asia has not come true. Instead, as of now, Northeast Asia is full of uncertainties and growing nationalism.

The optimism that Northeast Asian countries will become more cooperative after achieving remarkable economic growth is being replaced by the pessimism that they will become more competitive and confrontational with their growing national powers. It is uncertain whether the rise of China will aspire to contend with the United States in the decades ahead. In the meantime, different from the past when Japan was maintaining a low profile behind the security support from the United States, Japan is forthcoming to compete with China. North Korea is aggressively pursuing its nuclear weapons program, either to hedge against the uncertainties surrounding the Pyongyang's future, or to deter the plausible US preemption. Indeed, North Korea's nuclear test conducted in October 2006 caused policies of countries in Northeast Asia to be tougher than earlier toward North Korea.

In this backdrop, South Korea has been pursuing the policy of peace and prosperity on the Korean Peninsula and in Northeast Asia. However, Seoul's policy of peace and prosperity on the Korean Peninsula and in Northeast Asia has not achieved its policy goals, because the North Korean nuclear issue has been blocking the progress of the peace and prosperity policy. In fact, the Korean Peninsula has been under heightened tension due to North Korea's nuclear ambition, in addition to the North's conventional military threats. North Korea's nuclear issue has been out for almost two decades. The September 2005 Joint Statement of the six-party Talks is the third agreement, the first two being the 1992 Denuclearization Agreements of the Korean Peninsula between North and South Korea and the 1994 Geneva Agreed Framework between Washington and Pyongyang. Since the two earlier agreements were not faithfully observed by Pyongyang, whether the third one will unfold successfully remains to be seen.

In spite of the uncertainties surrounding the Korean Peninsula and Northeast Asia, the September 2005 Joint Statement of six-party Talks cast hope for participating countries in six-party Talks though the hope remained temporary. In fact, the six-party Talks has been stalling in the midst of disputes on issues of North Korea's counterfeiting and human rights between Washington and Pyongyang. The six-party talks were brought to a halt for more than a year before they resumed on December 2006. Nonetheless, six countries returned to the talks, after North Korea conducted a nuclear test, because it is impossible to resolve the North Korea's nuclear issue without negotiation.

In this sense, we can hardly discount the values of the September 2005 Joint Statement because it is the first agreement where six countries (South Korea, North Korea, the United States, China, Japan, and Russia) in Northeast Asia, for the first time in their history, participated to resolve the most serious security issue -North Korea's nuclear issue. The Joint Statement also becomes a reference point for the six counties to resume and make progress in order to resolve North Korea's nuclear issue completely. Though broad in its scope and nature, the Joint Statement addresses problems that North Korea and the other five countries have been facing for a long time. In essence, it commits North Korea to abandon nuclear weapons and programs, while the other parties commit themselves to provide security assurances and economic assistances to North Korea.

It remains to be seen whether North Korea will abandon its nuclear programs and weapons completely and verifiably, thereby irrevocably rendering North Korea non-nuclear for ever. This chapter addresses how to make the six-party talks a success in not only resolving North Korea's nuclear issue but also promoting security cooperation among six countries in Northeast Asia. This study also intends to address how to reduce military tension and build a durable peace on the Korean Peninsula. Lastly, this chapter will explore ways to promote multilateral security cooperation in Northeast Asia after having resolved the North Korea's nuclear issue and reduced tension on the Korean peninsula in the future.

Theoretical Exploration of Multilateral Cooperation for Northeast Asia

The integration theory tells us that the more contacts and exchanges nations have, the more cooperative and integrative their relationship will become. Social ignorance and social contact theories also view that the more information each has of the other and the greater number of contacts they make between the two, the less fear and confrontation they will have of each other. Likewise, the functionalist interpretation states that the outcome of political integration stems from economic integration of member states in a regional community where each country values common interests more than self interest or power.[1] The functionalists use the example of the European Union to prove the validity of their theory. They argue that economic exchanges and cooperation beyond national borders brought European countries together, ultimately leading to the European Union. They also argue that the German phrase "*Wandel durch Handel*" (change through trade) resulted in not only the German unification, but also accelerated the European Union. Despite merits in the functionalists' views,

1 David Mitrany, *A Working Peace System* (Chicago: Quadrangle Books, 1966), pp. 70-75.

functionalism has undergone harsh debate, mainly from the neofunctionalists who have a common basis on their theory of integration.

Neofunctionalism argues that the European integration was not the natural result of economic integration, but rather the result of the artificial efforts through supranational institutions of the European Union. Neofunctionalists discovered the weakness of functionalism in the slow progress of the European Coal and Steel Community (ECSC). They argued that progress in ECSC and other European supranational integration was made possible through agreed results from intergovernmental conferences in order to expand the mandate of EU legislation into new and more diverse areas. However, neofunctionalism finds its logic weak in that it assumes a natural decline in importance of nationalism and nation-states as the integration process unfolds. Neofunctionalists found the example of ECSC to explain the role and contribution of supranational institutions with higher authority than nation states to influence the acceleration of the international integration.[2] The existence of a higher authority is shown in the case of ECSC, whose major mission was to produce and manage coal and steel jointly beyond the monopolistic possession of one nation state.

The theory of intergovernmentalism came out of critiques of neofunctionalism. Intergovernmentalism bases its theory on the assumption that nation states or governments are making efforts to appoint their people to the international organizations and agencies to foster integration while they keep their sovereignty and loyalty of their appointees. Intergovernmentalists differ from neofunctionalists in that regional integration is neither a natural outgrowth of economic integration nor the role of supranational agencies which take over the loyalty and sovereignty of nation states voluntarily.[3] Independent appointees or elected representatives of member states and governments have solely advisory or implementation functions that their nation states and governments delegate. Intergovernmentalism uses the case of the European integration to show that governments control the level and speed of European integration.

The theory of constructivism in international relations most concerns itself with the role of ideas in shaping the international system.[4] By "ideas" constructivists refer to the goals, threats, fears, identities, and other elements of perceived reality that influence states and non-state actors within the international system. Constructivists believe that these ideational factors can often have far-reaching effects, and that they can trump materialistic power concerns. Therefore, there must be perceptions at work in shaping international outcomes. As such, international relations are constituted by exchanges of ideas and identities shared among leaders and civic groups in their countries. Constructivists find the success story of the European integration in the existence of shared identity, social consciousness, and information among European leaders and civic groups that have facilitated its integration.

2 Ernst B. Haas, *Tangles of Hopes: American Commitment and World Order* (Englewood Cliffs: Prince-Hall, 1969), pp. 10-12.

3 Robert Koehane and Stanley Hoffmann, eds., *The New European Community: Decision-Making and Institutional Change* (Boulder: Westview Press, 1991).

4 <http://en.wikipedia.org/wiki/International_relations_theory>

The theories mentioned above shed light on not only on the explanation of how countries could achieve the European integration, but they also provide insight as to how countries in Northeast Asia can achieve regional cooperation, if not entirely regional integration. Functionalists and Neofunctionalists are helpful in explaining factors for economic integration but fall short of explaining how Europe could achieve security cooperation that finally led to the European security community. The European security community was made possible by the inception of the multilateral Conference on Security and Cooperation in Europe (CSCE). There are three reasons for the making of the multilateral security cooperation process, so called, the Helsinki Process: (1) European countries, East or West, the United States and Canada, and the former Soviet Union made compromises on their respective national security interests to form the multilateral security cooperation process; (2) The Helsinki process pursued resolutions to political and security issues, economic and technology issues, and human rights issues at one forum, and; (3) The United States and the former Soviet Union tried to resolve their nuclear and conventional arms race in separate channels and they pursued arms control through negotiations to provide favorable environments for the progress in the Helsinki Process. This last reason is applicable to the Korean Peninsula and Northeast Asia in a broad perspective.

Resolving North Korea's Nuclear Issue at the Six-Party Talks

North Korea's nuclear issue has become a major blockade to the progress in peace and prosperity on the Korean peninsula and in Northeast Asia the two decades since the end of the Cold War. In February 2005, Pyongyang boldly announced that it had possession of nuclear weapons. Following that statement, in July 2006, North Korea conducted missile test launches of various kinds and conducted, in October 2006, a nuclear test whose blast yield is estimated to range between 0.2 and 1.0 KT.[5]

A shock wave went through the world despite the fact that Pyongyang had been playing the nuclear card for years. Before its official announcement of possession of nuclear weapons and subsequent nuclear test, much was left for speculation on how capable North Korea's nuclear weapons program was. Despite their official announcement of possession of nuclear weapons in February 2005 and nuclear test of October 2006, intelligence sources have disclosed conflicting views on just how many nuclear weapons North Korea may have possessed. Before Pyongyang's official announcement of nuclear weapon possession, the intelligence estimates ranged from the mere possession of 10 to 15 kilograms of plutonium to one or two crude bombs.[6]

5 Daniel A. Pinkston and Shin Sungtack, "North Korea Likely to Conduct Second Nuclear Test," Center for Nonproliferation Studies, Monterey Institute of International Studies, January 8, 2007. Siegfried S. Hecker, "Report on North Korean Nuclear Program," Center for International Security and Cooperation, Stanford University, November 15, 2006.

6 The South Korean government noted that it was assessed that North Korea may have produced one or two nuclear weapons. The ROK Ministry of National Defense, *Defense White Paper 2004*, p. 44. Director of the U.S. Central Intelligence Agency also mentioned that North Korea is assessed to possess one or two nuclear weapons. George J. Tenet, "Worldwide Threat:

As time went on, the estimation was made closer to four to thirteen nuclear bombs in North Korean hands.[7] Contrary to the general expectation that North Korea would refrain from developing nuclear weapons in light of Washington's punitive attack on Iraq, North Korea seemed to have been emboldened, admitting to its possession and plans to develop its nuclear weapons further. A North Korean Spokesman of Foreign Ministry stated that North Korea made a spurt to strengthen nuclear deterrent capabilities.[8] Pyongyang has disclosed that it has been developing nuclear weapons not only for deterrent purposes, but also because of its economic stringency that cannot provide resources to compete with South Korea and the United States in terms of military balance.[9]

More alarming is North Korea's continuous development of a ballistic missile program. Since the mid-1970s, North Korea has been developing its missile capabilities. Starting from reverse engineering of SCUD-B and SCUD-C missiles, North Korea's missile variety and range has grown exponentially. So far, North Korea produced and deployed 600 SCUD-B/-C, 150 Nodong, tens of Taepo-dong-1, and is arduously developing long-range ICBMs, which have the potential to threaten the North American continent effectively with a nuclear sized payload.[10] North Korea's missile development pattern is unique because it is used to create the capability to deploy missiles in the battlefields after one or two rounds of missile tests. The July 2006 missile tests demonstrated how seriously North Korea has been developing a variety of missile programs. This event exceeded China's limit of patience, thereby causing China to agree to the United Nations Resolution 1695, the contents of which was to penalize North Korea unanimously in the United Nations Security Council.

North Korea's Accusations and Motivations for Nuclear Weapons

Since the advent of the George W. Bush administration, North Korea has stepped up its anti-U.S. propaganda, saying that North Korea's policy stance to strengthen its deterrent power was a wise decision to cope with President Bush's hard-line policy toward Pyongyang. Immediately after Bush's victory in the presidential election, North Korea emphasized the obligation of the United States' to observe and abide by the Geneva Agreed Framework and the non-hostile Joint Statement

---

Converging Dangers in a Post-911 World," Testimony of Director of Central Intelligence Agency before the Senate Select Committee on Intelligence, February 6, 2002.

7 Dr. Wolfstahl's estimate is the biggest in telling us that North Korea has, at maximum, potential to make 13 nuclear bombs. Jon B. Wolfstal, online at <http://www.ceip.org>. Jon B. Wolfstal, "Nuclear Threat Reduction in North Korea," Discussion Paper of the Joint CSIS-Carnegie Project on Korean Threat Reduction, September 20, 2004. The US presidential candidate of the Democratic Party, John Kerry mentioned that North Korea may have possessed four to seven nuclear bombs, on which he argued to indicate a total failure on the part of the President Bush's North Korea policy. *International Herald Tribune*, October 2-3, 2004, p. 5.

8 The DPRK Foreign Ministry Spokesman Statement, June 18, 2003.

9 Commentary of the North Korean Central News Agency, June 9, 2003.

10 IISS, North Korea's Weapons Programmes: A Net Assessment, An IISS Strategic Dossier, January 2004, pp. 74-83.

between Washington and Pyongyang which was agreed in October 2000 between the North Korean Marshall, Cho Myong-rok, and the U.S. Secretary of State, Madeline Albright. North Korea was in fear of the Bush Administration, which criticized the Clinton Administration for allowing North Korea to cheat on the Geneva Agreed Framework.

It appears that North Korea's sense of insecurity was aggravated by President Bush's speech naming the "Axis of Evil," and subsequent adoption of new nuclear posture review in 2002. Pyongyang showed a fanatical response to those initiatives and felt vulnerable against Washington's threat of war or use of nuclear weapons. Out of its anxiety that its economic stringency cannot provide resources to compete with South Korea and the United States in terms of conventional military balance, the North Korean leader may have come to the conclusion that his regime should pursue weapons of mass destruction, no matter what incentives the external world provides to North Korea. For the security of its regime, North Korea paradoxically believes that it requires nuclear weapons.

According to a report produced by the U.S. Council on Foreign Relations,[11] Pyongyang's attitude toward nuclear weapons has hardened over time, especially since February 2003. The report writes that Pyongyang's appeals for negotiations have been mixed with an increasing number of statements defending its right to produce nuclear weapons. Thus, it may or may not be open to serious negotiations. In the aftermath of the war on Iraq, Pyongyang may have concluded that deterrence is more easily attainable through nuclear weapons than through negotiations with the United States. The other side of the coin was that the more Washington regarded North Korea as "an evil axis," which Washington added to a "rogue state" that the Clinton Administration branded,[12] the more often Pyongyang insisted upon having direct talks with Washington to reduce its sense of insecurity. In other words, North Korea's fear of collapse was manipulated by Washington.[13]

Ironically, Pyongyang insisted upon concluding a non-aggression pact with the United States as a *quid pro quo* to its abandonment of its nuclear weapons program. Therefore, the September 2005 Joint Statement provided North Korea with the U.S.'s security assurances by stating that the United States has no nuclear weapons on the Korean Peninsula and has no intention to attack or invade the DPRK with nuclear or conventional weapons. In addition, Pyongyang seems to be interested in normalizing relations with Washington, including Washington's lifting of economic sanctions against the DPRK, plus energy and financial compensation. Pyongyang is nervously and dangerously playing a nuclear game with a risk-seeking mindset while also seeking a negotiated settlement. Yet it remains uncertain what price North Korea is truly willing to pay in return for the benefits it is aspiring to extract from the United States.

---

11 Council on Foreign Relations, *Meeting the North Korean Nuclear Challenge*, May 19, 2003, online at <www.cfr.org/publication_print.phpid>.

12 Leon V. Sigal, "A Rogue by Any Other Name," *Foreign Service Journal*, October 2005, pp. 37-44.

13

The United States' Strategy and Policy

There was a big shift in the U.S. policy toward North Korea from the Clinton to the Bush administrations. The latter adopted a policy toward North Korea to resolve the whole panoply of North Korea's threats—ranging from the nuclear, missile, and conventional variety—which differed greatly from the former's policy, which preferred to resolve the nuclear and missile issues in succession. Criticizing the previous administration for giving too much for too little, President Bush made it clear that Washington would not give anything on account of North Korea's violation of the non-nuclear North Korea agreements.

According to Washington, Pyongyang's nuclear weapons destabilize peace and security on the Korean peninsula and in Northeast Asia, as well as endanger the international nonproliferation regime. Japan, South Korea, and Taiwan could begin to develop nuclear weapons to deter and defend themselves from a nuclear-armed North. Pyongyang would probably try to widen the chasm between Seoul and Washington at the moment when the two allies may have different priorities in dealing with Pyongyang. Mysteriously enough, Washington imposed financial sanctions on the DPRK because of its counterfeiting and violation of human rights in the name of ending tyranny in North Korea. Now that economic sanctions seem to be more effective in dissuading North Korea from sticking to its nuclear weapons program, Washington has been pursuing tougher measures, including proliferation security initiatives and UN Sanction, both before and after North Korea's nuclear test.

However, Washington showed signs of accommodating North Korea's long-held request to have direct talks with Washington after North Korea's nuclear test and brinkmanship diplomacy. As the Democratic Party won the majority in the U.S. House and Senate through the mid-term elections, the Bush Administration needed to accommodate the Democratic Party's request to be serious about having direct talks with North Korea.[14] In Berlin, Christopher Hill, head of the U.S. delegation to the six-party Talks met with Kim Kye-gwan, head of the DPRK delegation to the six-party Talks and the meeting paved way for the February 2007 six-party Talks in order to take initial measures to implement the September 2005 Statement. Overall, it is uncertain how the United States will conduct a give and take approach and with what combination of incentives and disincentives in order to extract concessions from North Korea.

China's Strategy and Policy

China has traditionally supported a non-nuclear Korean peninsula. The United States, South Korea, and Japan specifically thanked China for hosting the six-party talks in Beijing and they underscored that China's cooperation would play a key role in the peaceful resolution of the North Korean nuclear issue. Nevertheless, Beijing seemed to oppose Washington's hard-line policy toward North Korea. Beijing opposed even

14 Donald Kirk, "The Korean Bomb Still Ticking," Asia Times, November 11, 2006. online at <http://www.atimes.com>.

further the U.S. financial sanctions on Pyongyang because they could provide North Korea's excuse to boycott the six-party Talks.

The Chinese government used to disagree with Washington's hard line policy toward North Korea because it contended that the U.S. hard line policy may cause North Korea to accelerate its nuclear program instead of abandoning it. Although China played a mediating role in persuading North Korea to abandon nuclear programs and in persuading the United States and Japan to provide security assurances and economic incentives in return for North Korea's concession, remarkable progress in the six-party talks has not been made except for the September 2005 Statement. It was uncertain how far China would go along with the United States to push North Korea to abandon its nuclear weapon program until North Korea conducted a nuclear test.

After North Korea's nuclear test, China condemned North Korea in an unprecedented tone for being flagrant and stubbornly defiant.[15] China is known to have conveyed a strong warning to the North Korean authority to suspend any more nuclear tests and reconsider the China-North Korea relations. Urged by the UN resolution 1718 to use its economic assistance to press North Korea, it was seen that China was seriously reconsidering its relationship with North Korea. Chinese experts resonated with the international community to condemn and take a tough stance against a nuclear North Korea. Within the Chinese decision making process, it was known that China was considering a shift in the China-North Korean relationship, from the traditional and special allied relationship to the national interest-centered and normal relationship.[16] By the end of 2006, the Chinese government clearly stated that North Korea's missile test launch and nuclear test pose serious destabilizing challenges to the Korean Peninsula and Northeast Asia.[17] Undoubtedly, China's strong stance on North Korea worked to bring North Korea back to the six-party talks.

South Korea's Position and Policy

Seoul made it clear that it would not tolerate North Korea's nuclear weapons and at the same time, the issue should be resolved peacefully through dialogue. However, South Korea showed a difference in style and emphasis in its approach to the North Korean nuclear issue than the United States. As the six-party talks stalled, South Korean government came to the conclusion that there is no option but to improve inter-Korean exchanges and cooperation within some boundaries, of which the bulk is to continue three inter-Korean projects—the Mountain Geumgang tourism project, the Gaesong Industrial Complex, and the reconnection of railroads and roads. Given

15 The PRC Ambassador to the United Nations, Wang Guangya said so in the United Nations. October 9, 2006. Requote from John J. Tkacik, Jr., "A New Tack for China after North Korea's Nuclear Test?" Policy Forum Online 06-91A, October 26, 2006, Nautilus Institute, online at <http://www.nautilus.org/fora/security/0691Tkacik.html>.

16 The author's interview with Dr. Donqiu Lee who visited Korea on December 1, 2006.

17 The PRC Government, China's National Defense 2006, December 2006.

the U.S. new policy of imposing financial sanctions and the human rights standards on Pyongyang, Seoul seems to be at odds with Washington because those sanctions do not produce any progress in the six-party talks and inter-Korean reconciliation and cooperation.

However, South Korea's policy toward North Korea faced grave challenges after North Korea's missile test launches and nuclear test. The South Korean public opinion requested the government to reconsider its North Korea policy and join in the UN-led sanctions. The Roh Moo-hyun government temporarily expressed their will to reconsider its accommodation policy toward North Korea. Urged by the United States, the South Korean government took into consideration whether or not to join fully in the US-led proliferation security initiative to press North Korea and whether or not to suspend economic aid to North Korea. Above all, the South Korean bottom line is to continue the six-party talks to resolve the North Korean nuclear issue with willingness to contribute to the economic and energy incentives to be provided to North Korea in return for North Korea's abandonment of its nuclear programs.

Japan's Policy and Strategy

Japan also strongly supports a non-nuclear Korean peninsula. Japan clearly sees a threat from North Korea due to its nuclear and missile programs.[18] The reason why Japan is conducting joint research and development of a missile defense system with the United States is North Korea's continuous development of missiles and nuclear weapon programs. On North Korea's nuclear and abduction issues, Japan is taking a tough stance. Japan's approach is closer to the United States' than to South Korea's. Japan sticks to the rule that North Korea's nuclear weapons program should be eliminated in a complete, verifiable, and irreversible manner, and that tougher measures are necessary if North Korea does not give up its nuclear programs during the six-party talks.

However, Japan's apparent priority in the six-party talks was laid upon the resolution of the abduction issue rather than the nuclear issue. Japan is following the U.S. policy priority in relation to the proliferation security initiative. In that connection, Japan agrees with the United States on the necessity of economic sanctions and human rights that Washington has pursued recently. Japan took the lead, along with France in the United Nations to pass the UN resolution 1718, to impose sanctions on North Korea. Furthermore, Japan put financial sanctions on North Korea immediately after the passage of the UN resolution.

Russia's Policy and Strategy

Russia does not want North Korea to develop nuclear weapons. However, Russia seems to lean toward a regional approach in which it can play a role. Russia has also shown a keen interest in using its supply of natural gas and connection of the Trans-Siberian Railway with the Trans-Korean Railway in a way that could induce North Korea to abandon its nuclear weapons program. However, Russia is not willing to

18 Japan's Self-Defense Agency, *Defense White Paper 2006*, October 2006.

pay any costs for economic incentives to be provided to North Korea. Russia has not supported U.S.-led economic sanctions and human rights policy toward North Korea, even after the North Korea's nuclear test. Moscow is more interested in a negotiated settlement of North Korea's nuclear issue.

Six-Party Talks and Beyond

After the eruption of the second nuclear crisis in October 2002 and North Korea's second announcement to pull out from the Nuclear Nonproliferation Treaty, South Korea, the United States, China, Japan, and Russia laid out a common strategy to bring North Korea to the six-party talks in order to change North Korea's status to non-nuclear. Though the six-party talks could take a long time to resolve the North Korean nuclear issue due to too many cooks spoiling the broth, it provides a better chance of negotiated settlement as long as North Korea and the United States are willing to negotiate within this forum.

First, six-party Talks are better than two- or three-party talks not only because North Korea's nuclear issue is regional, but because it stands a better chance for a negotiated settlement by making a big pie of incentives to be provided to North Korea.[19] Second, North Korea could find a face-saving solution more easily in the multilateral talks than in impracticable bilateral talks with the United States. As for the nuclear inspection regime on North Korean nuclear programs and facilities, Pyongyang will be less humiliated and forceful in its reaction when compared to its possible reaction to an inspection regime imposed on North Korea by either the United Nations Security Council or the United States. Third, the six-party talks were left as the only feasible negotiation venue, not only because the United States strongly opposed any direct talks with North Korea at the risk of rewarding North Korea's bad behavior, but also the United States wanted China to play a mediating and constructive role on the North Korean nuclear issue. In retrospect, the bilateral talks between South Korea and North Korea in 1991 and 1992, followed by the U.S.-DPRK talks, have not succeeded in halting and dismantling North Korea's nuclear programs. Fourth, five countries (except for North Korea) can take a more integrative approach by combining requirements from North Korea and their provisions to North Korea. Along this line, a CSIS policy report came out with the title of, *Cooperative Threat Reduction and North Korea*.[20] Fifth, six parties can ensure that steps to resolve the North Korean nuclear problem will be so incremental and reciprocal that any military option could only come after all other steps have been thoroughly taken.

The first round of six-party Talks was held in August 2003 in Beijing. Since then, five rounds of six-party Talks convened on and off after long intervals in between, thereby the fifth round of talks were being held in February 2007. Due to big differences in strategy and policy of the six countries, there were only discussions

---

19 Yong-Sup Han, *Peace and Arms Control on the Korean Peninsula* (Seoul: Kyungnam University Press, 2005), pp. 29-39.

20 Joel S. Wit, Jon Wolfsthal, and Choong-suk Oh, *The six-party Talks and Beyond: Cooperative Threat Reduction and North Korea* (Washington, D.C.: Center for Strategic and International Studies, 2005).

until September 19, 2005. The six countries managed to produce the September 19, 2005 Joint Statement to make North Korea abandon its nuclear weapons and related programs. The September 2005 Joint Statement calls for North Korea to commit to "abandoning all nuclear weapons and existing nuclear programs."

Although North Korea claimed the right for peaceful uses of nuclear energy, including immediate provision of light-water reactors, the other five countries maintained that the six parties have agreed to discuss the issue "at an appropriate time," which meant some time after North Korea has been verified to have dismantled its nuclear weapons and programs. To facilitate the progress in the six-party Talks, South Korea promised to provide 200 Megawatts of electricity as a grand deal to compensate for North Korea's abandonment of nuclear reactors to be constructed and suspended in Shinpo, North Korea.

The Joint Statement also commits "the directly related parties" to negotiate a permanent peace regime on the Korean Peninsula at a separate forum. During the 1990s, Four Party Talks among South and North Korea, the United States, and China had been held in Geneva. The talks collapsed due to North Korea's refusal to continue the Four Party Talks unless the four countries agreed to discuss withdrawal of US forces from Korea. It is not still clear whether Pyongyang will accept South Korea as a legitimate partner in peace talks, because so far it has denied South Korea's participation in the peace talks. Instead, North Korea used to insist that it and the United States be legitimate partners for peace talks on the Korean Peninsula. Pyongyang's primary goal concerning the peace treaty used to isolate South Korea and undermine the ROK-US security alliance.

However, the six-party talks were almost derailed due to North Korea's insistence upon release of suspended bank accounts in Banc Delta Asia by the United States Treasury Department before the resumption of nuclear talks. Immediately after the September 2005 Joint Statement, Washington put financial sanctions on North Korean bank accounts in Banc Delta Asia based in Macao, China. This was interpreted as a conspiracy to strangle the Kim Jong-il regime to collapse. Suspecting the U.S. policy intentions to bring down the Kim Jong-il regime in the name of ending tyranny as part of transformational diplomacy, North Korea resisted strongly against the US sanctions. Then North Korea challenged the United States to choose either bilateral DPRK-US talks or a nuclear armed North Korea. Finally, North Korea conducted a nuclear test and the United Nations Security Council passed the UN resolution 1718 to strengthen sanctions on North Korea: cargo inspections, arms embargo, asset freezes, and travel ban.[21] However, the international community, in particular China and South Korea, was not following US-led sanctions initiatives, whereas the United States and Tokyo took tough measures on North Korea. Due to Washington's primary obsession with Iraqi and Iranian matters, Washington had to begin to find out ways to bring North Korea back to the six-party talks. At the same time, Pyongyang had no other choice but to seek ways to return to the six-party Talks because Kim Jong-il was strongly warned by the Chinese authority not to repeat recalcitrant behavior

21 Larry A. Niksch, *North Korea's Nuclear Weapons Development and Diplomacy*, CRS Report for Congress, The U.S. Congressional Research Service, updated January 3, 2007.

on the one hand and suffered the effects of sanctions on North Korea on the other hand.

In addition, President Bush mentioned that the United States would be interested in declaring the end of the Korean War if North Korea faithfully abided by the non-nuclear principle. This concession is meant to address the need for a peace regime on the Korean Peninsula. As for a modality to discuss a peace regime in the future, South Korea should participate as a legitimate and legal partner. Nevertheless, a peace building process will take a long time because of a great deal of differences in the approaches of the four countries such as South Korea, North Korea, the United States, and China. The Korean peace process, in essence, is comprehensive because it entails tension and threat reduction on both sides of the two Koreas and the U.S. forces in Korea, in return for security guarantees to North Korea and provision of economic assistance to North Korea.

The success of the six-party talks hinges on who does what and when through what kind of sequential implementation. The next six-party talks will have to address the issues of specifying plans, measures, and procedures to implement the September 2005 Statement. This will include resolutions that: North Korea should abandon all nuclear programs, and promise to dismantle its nuclear weapons and facilities; The six parties should agree to impose verifications on North Korea and North Korea should accept such verifications as agreed upon by the six parties; The other five parties should agree what economic incentives including energy assistances they will provide to North Korea, on condition that North Korea abandon and dismantle its nuclear weapons and programs and North Korea should accept the agreements; North Korea's return to the NPT and verification of North Korea's compliance to a non-nuclear North Korea will be following phases, starting from freeze and shutdown to total elimination of programs and capabilities as agreed by the six parties; Economic assistance to North Korea should be multilateral and conditional upon North Korea's progress in abandoning and dismantling its nuclear weapons and programs; The six-party Talks should continue until and after they achieve a non-nuclear Korean peninsula and are turned into a regional security cooperation mechanism.

Building a Peace Regime on the Korean Peninsula

Although North Korea's nuclear weapons program is a number one priority issue at present, North Korea's conventional military threat is second to none in relation to peace and security on the peninsula, as well as in Northeast Asia. In spite of its economic difficulties, North Korea's conventional military threats continue to affect security situations negatively. North Korea's relentless pursuit of weapons of mass destruction aggravates the overall military balance on the Korean Peninsula. Therefore, South Korea still feels threatened by North Korea's military threats.

South Korea's Sunshine Policy and Peace and Prosperity Policy have not been successful in reducing military tension on the Korean peninsula. It is, however, notable that Seoul's North Korea policy has been conducive to ameliorating the political and economic relationship between North and South Korea. Furthermore,

Pyongyang is interested in continuing exchanges and cooperation in the economic, social, and cultural areas. Seoul's approach to conventional arms control so far has been to discuss easy matters first and move to difficult measures later on: starting from military confidence building measures, to constraint measures, and finally to arms reduction measures. The approach has not achieved a substantial progress because North Korea has insisted upon arms reduction, in particular, reduction or withdrawal of U.S. forces in Korea.

The reluctance of South Korea and the United States to talk about arms reduction can be attributed partly to their lack of trust in North Korea, and partly to their concern that arms reduction talks would accelerate a U.S. pullout more than corresponding arms reduction on the North Korean side, which would also be hard to verify. Such reluctance also stems from the military's preference of the *status quo ante* instead of force reduction. North Korea may also avoid arms reduction for fear that the result might bring political and social instabilities.

However, it is time for us to design conventional arms control measures through a comprehensive approach, rather than focusing on military confidence building with North Korea *per se*. The comprehensive approach represents the best solution to negotiated tension reduction on the Korean peninsula. It creates a formula to satisfy the two Koreas and the United States. It also prevents the North Korean military from abusing economic assistance that the country receives from South Korea and the international community. By addressing North Korea's threat unequivocally and entirely, this approach has a better chance of ensuring that a deal is struck.

Therefore, South Korea should use the improved inter-Korean atmosphere to probe North Korea's willingness to agree to a larger deal of military confidence building and conventional arms control, in addition to the WMD and missile issues. This will not only help rebuild North Korea's economy but also thaw the military confrontation on the peninsula. Without such a grand bargain that trades, in essence, South Korea's and other countries' economic assistances for North Korea's pullbacks of its forward deployed troops along the Demilitarized Zones,[22] it is difficult for the South Korean government to attain broad political support for its reconciliation and cooperation policy toward the North, from inside South Korea let alone from the external world. For this option to succeed, South Korea should regularize the military meetings with North Korea, and then invite the United States to talk about arms reduction in tripartite talks later. In order to talk about conventional military issues, tripartite talks offer the best means for discussing conventional arms control in Korea.

The desperate economic situation that North Korea finds itself in and the leadership's motivation to engage South Korea opens up a window of opportunity for military confidence building and conventional arms control. If South Korea and the United States design an articulated joint strategy to induce North Korea to entertain arms control talks, the chances for arms control will increase a great deal. Arms control is a required process for South Korea to build a peace regime on the Korean Peninsula. When the North Korean nuclear issue is at a deadlock, contemplating

22 Michael O'Hanlon and Mike M. Mochizuki, *Crisis on the Korean Peninsula: How to Deal with a Nuclear North Korea*, 2004.

how to start an arms control process on the Korean Peninsula is worth highlighting because it can facilitate the resolution of the nuclear issue directly or indirectly.

Regional Security Cooperation in Northeast Asia

It follows that regional security cooperation among countries in Northeast Asia will enhance a durable peace on the Korean peninsula. Likewise, a durable peace regime on the Korean Peninsula will also reinforce regional security cooperation.[23] A decade ago, and to a lesser extent still now, the concept of Northeast Asian security and the thought of building an indigenous security framework confronted considerable skepticism, both in Northeast Asia and in the West, because of Asia's lack of common threats, interests, cultures, political and economic ideologies and systems, the region's geographical dispersion, historical animosities, etc.[24]

Northeast Asian countries, except for North Korea, have showed remarkable economic growth, which will require countries to protect and promote security and prosperity individually and collectively. As security interests grow with growing national wealth, security as a viable concept becomes more compelling.[25] It is true that rising nationalism based on historical animosities still hinders multilateral security cooperation commensurate with multilateral economic cooperation. Nevertheless, if North Korea's nuclear issue can be resolved through the six-party Talks, the speed for multilateral security cooperation within Northeast Asia will be accelerated. Ideas to build on the six-party Talks will blossom in order to found the Northeast Asia Security and Cooperation Conference.[26] Northeast Asians can also capitalize upon achievements made through the ASEAN Regional Forum, Northeast Asia Cooperation Dialogue, and the APEC, so as to navigate themselves through tempests like WMD proliferation and international terrorism. They will be able to begin charting a new security concept that will include all Northeast Asians and realize a collective quest for peace, security, and economic growth.

Northeast Asia can not be an exception from the aspiration for peace, security, and sustained economic growth. Until now, security in Northeast Asia has been dependent upon a set of bilateral alliances. However, Northeast Asian countries desperately need a multilateral security cooperation mechanism for three fundamental reasons. First, Northeast Asian countries should foster security cooperation in light

---

23 Yong-Sup Han and Kyung-Yung Chung, "Promoting Multilateral Security Cooperation in Northeast Asia: Findings of the Northeast Asia Security Policy Forum and Its Future," *The Korean Journal of Security Affairs*, December 2006, vol. 11, no. 2, pp. 7-30.

24 Thomas U. Berger, "Power and Purpose in Pacific East Asia: A Constructivist Interpretation," G. John Ikenberry and Michael Mastanduno, ed., *International Relations Theory and the Asia-Pacific* (New York: Columbia University Press, 2003) pp. 387-19.

25 Yong-Sup Han, "Implications of the European Success of Multilateral Security Cooperation for Peace in East Asia and Northeast Asia," *The Korean Journal of Security Affairs*, June 2006, vol. 11, no. 1, pp. 83-102.

26 James Goodby, "The six-party Talks: Opportunity or Obstacle?" Jeju Development Institute and East Asia Foundation, *Building a Northeast Asian Community: Toward Peace and Prosperity*, Yonsei University Press, 2006, pp. 144-72.

of comprehensive security to further develop their economic interactions and deepen economic interdependence. Without security cooperation, economic interaction will face risks stemming from heated competition to search natural resources and energy, deal with environmental problems, natural disasters, diseases, attacks on information networks, drug trafficking, illegal migration, terrorism, not to mention the lingering suspicion of neighboring countries.

Second, there will be an increase in arms competition among countries in the region. Northeast Asia is the only region where major powers are competing to modernize armed forces in an unprecedented speed and scale because of their growing economies and hegemonic rivalry. Without cooperative security in place, countries in the region will avoid openness and transparency in their defense policy and arms build-up. Sooner or later, they will face the security dilemma that European countries had faced in the 1970s and 1980s.

Third, without a multilateral security cooperation framework, they cannot reduce antagonism stemming from the historic rivalries, animosities, growing nationalism, assertions, territorial claims, and history book fights, etc. During the first half of 2000s, political leaders of Northeast Asian countries showed a propensity to extract domestic political support by arousing national sentiments against neighboring countries. Political and security leaders in Northeast Asia are advised to gather at one forum on a regular basis in order to enhance mutual understanding and confidence, so as to create a shared vision of the Northeast Asian community building process. By doing so, they can lead their own people to implant a regional sense of mind and drive away their ethnocentrism and narrow mindedness.

Therefore, it is critical that Northeast Asian countries collectively resolve the North Korean nuclear issue before entering any serious discussion about the multilateral security cooperation. However, it is equally important that Seoul, Pyongyang, Beijing, Washington, Tokyo, and Moscow pursue regional security talks to help reduce military tensions on the Korean Peninsula and to resolve the North Korean nuclear issue through the six-party Talks.

Recently, it can be discovered that there are limitations in functionalist and neofunctionalist approaches to regional cooperation in Northeast Asia, mainly through economic exchanges and cooperation. In Northeast Asia, it is necessary for nation states and governments to take the lead in fostering regional cooperation by building the multilateral institutions and dispersing regional identity and values. Northeast Asia requires introduction of a new vision and ideas to formulate a common understanding and identity of the region in pushing toward a regional cooperation, even if the regional integration remains a distant goal. Hence, it is necessary that the constructivist theory is able to play its role in creating the Northeast Asian community.

Index

Note: Bold page numbers indicate tables. Numbers in brackets preceded by *n* are footnote numbers.

ABC (Anything But Clinton) policy 152-154, 184
Abe, Shinzo 125, 126, 130
ACSA (Acquisition & Cross-Servicing Agreement) 116
Afghanistan 84, 92(*n*126)
Agreed Framework (1994) 1, 12-13, 16, 66-67, 68, 72-73, 84, 93, 101, 111, 137, 169-170, 191
  collapse of 112, 113, 170, 171, 182-186
  confidence building measures 169
  lessons from 157-159
  three key elements of 174-175
  see also KEDO
Agreed Statement (1993) 72
aid 3, 4-5, 37, 82, 114, 115, 137-138, 144
  see also energy aid; food/fertilizer aid; *and see under individual countries*
Al Qaeda 4, 31
Albright, Madeline 171, 196
Alexeyev, Alexander 139, 146
APG (Asia-Pacific Group on Money-Laundering) 25
appeasement 10, 13, 65, 69, 87
Argentina 179
Armitage Report (1999) 183-184
ASEAN Regional Forum 204
Asia 59, 77
Asia-Pacific region 97, 134
Asian Financial Crisis (1997) 181
Aso, Taro 125, 126
Australia 179
"Axis of Evil" 9, 11, 29, 47, 67, 68, 147, 153, 184-185, 196
  within US administration 54

ballistic missiles
  defence (BMD) 119, 124, 130, 199
  DPRK's capability 195
  tests *see* missile tests
Ban Ki-moon 17, 106
BDA (Banco Dalta Asia) 6, 22-25, 33, 35, 36-37, 38-39, 40, 73-74, 104, 121, 140, 157, 164
  agreement over 145
Beijing agreement (2007) 76, 79, 84, 87, 89, 93, 94
"benign neglect" policy 26
Bin Laden, Osama 31
BMD (ballistic missile defence) system 119, 124, 130, 199
Bolton, John R. 5, 69, 74, 84, 87, 91-92, 93
Bosworth, Stephen 187
brinkmanship 29, 49, 52, 59, 102, 151, 158-159, 160, 164, 166
Brookes, Peter 176, 183
burden sharing 173
Bush Doctrine 47, 49-50, 81
Bush, George Sr 66
Bush, George W. 4, 47, 53, 91-92, 106, 153, 171
  "axis of evil" speech (2002) 9, 11, 29, 47, 67, 68, 153, 184-185, 196
  see also United States (Bush administration)

Canada 179
Carlin, Robert 9
Carter, Ashton 89
Cato Institute 53
Cha, Victor 73, 84(*n*77), 92, 92(*n*130)
Cheney, Dick 27, 86, 152-153
Chile 179
China 1, 3, 7, 11, 30, 54-57, 60, 74, 95-109, 136, 159, 197-198
  aid to DPRK 4-5, 109, 138, 144
  Communist Party 96, 97, 98, 101, 105
  "cooperative security" policy 103
  and counterfeiting 25
  development in 105

Friendship Treaty (1961) 99-100, 99(*n*9), 109
as great power 95-96, 97, 108
Hu Jintao visits DPRK 104-105
Imperial period 167-168
interests 95-96
and Japan 95, 96, 97, 109, 129, 168
Kim Jong-Il visits 105
multilateralism of 97, 98, 108
and "new security concept" 97-98
nuclear deterrence of 4
as nuclear power 97
"peaceful emergence" strategy 98
reaction to missile/nuclear test 7, 36, 96, 97, 98, 99, 100, 106-107, 141, 142, 195, 198
regional role of 11, 95-96
and ROK 56-57, 95, 96, 98, 107, 109
and Russia 109, 134, 135, 146, 149
sanctions 32, 105, 106, 198
as security concern 191
in six-party talks 18-19, 20, 70, 82, 94, 101-104, 107-109
Joint Statement 103, 108
and TMD 97
trade with/investment in DPRK 96, 98, 106, 109
in trilateral talks (2003) 154-155
and UN 27, 100, 195
and US 1, 3, 10, 30, 36, 56-57, 58, 68, 77-79, 96, 98, 107-108, 109, 155, 161
divergent priorities of 79
hegemony of US 77-78, 99
Cho Myong-rok 196
Chosen Soren 122-123
Chung Dong-young 17
Cleveland, Paul 179
Clinton administration 12, 13, 66, 72, 101, 152, 153, 169-184
ACDA (Arms Control & Disarmament Agency) 173, 174
Congress 170, 172, 182-183
Defence Dept (DoD) 173, 175
Energy Dept (DoE) 173, 174
helicopter incident (1994) 174
Intelligence/Research Bureau 173
Korea Affairs, Office of 173, 174
mid-term elections 170, 172
"Clintonian appeasement" 65, 85, 93, 182, 184, 196
Cold War 45, 46-47, 50-51, 61, 91, 108, 168
"compellant threat" 83
"compensation for freezing" 138, 156, 161-162, 165, 166
"comprehensive dismantlement" 155-156
comprehensive security 62, 152
Confucian tradition 56
constructivism 193
"cooperative security" 103
Cotton, James 74(*n*35)
counterfeiting 22-25, 33, 38, 48, 82, 140-141, 164, 197
see also BDA
crisis escalation tactics 160-161, 163, 164
assessment of 165
CSCE (Conference on Security & Cooperation in Europe) 194
CVID (complete, verifiable and irreversible dismantlement) 2, 68, 72, 79, 93, 114, 137, 155, 166

Dai Bingguo 101, 136
democracy 58, 88
Denisov, Andrei 143
denuclearization road map 39-40, 41, **41-42**
diplomacy 9, 11, 20, 35, 65, 66, 87, 154-155, 164, 169, 172
Chinese 95, 96, 98, 101-102
and issue linkages 81
transformational 201
disablement 4, 5, 19, 20, 40
disclosure issues 8
DPRK *see* North Korea
drug trafficking 26, 71, 82, 92(*n*130)

"early harvest" 66
Eberstadt, Nicholas 80(*n*59), 85-86, 89
Elliott, Kimberley Ann 82
energy aid 4-5, 7, 8, 12-13, 37, 102, 114, 118
European Coal & Steel Community (ECSC) 193
European Union 179, 182, 192-193
exceptionalsim 83

face-saving 160, 163
family reunion project 30

fertilizer sanctions 32
 see also food/fertilizer aid
food shortages 51, 180
food/fertilizer aid 3, 4, 28, 30, 117, 144, 180
 misuse of 182-183
Four Party Talks 201
"freeze for compensation" 138, 156, 161-162, 165, 166
Frum, David 47
functionalism 192-193, 194

Gaesong Industrial Complex 3, 30, 198
Gallucci, Robert 172-173, 177
Geneva Agreed Framework *see* Agreed Framework
Germany, East/West 61
Gilman, Benjamin 184
Glaser, Daniel 24, 35
GNP (Grand National Party) 80
Green, Michael 183
GWOT (global war on terrorism) *see* war on terror

Hadley, Stephen J. 6
Han Son-ryol 28
Hathaway, Robert 85
"hawk engagement" 10, 66, 84(*n*77), 86-87, 88, 93-94
Helsinki Process 194
Henderson, Gregory 45
Hersh, Seymour 84(*n*81)
HEU (highly enriched uranium) 1, 2, 7, 8, 15, 19, 20, 39, 67-68, 112, 118, 136, 145, 153-154
 DPRK denies making 185
HFO (heavy fuel oil) 4-5, 7, 8, 12, 113, 115, 118, 136, 144, 169, 174-175
 payment/supply of 175
 power plants for 179
 suspicions surrounding 182
Hill, Christopher 6, 7-8, 18, 20, 21, 28, 35, 43, 72, 74, 93, 106
 and Kim Kye-gwan 36-38, 144, 145, 156, 163, 165, 197
 on money laundering 23
history books 205
"hostile neglect" 65
Hu Jintao 7, 23, 30, 96, 97, 101, 104, 164
Hubbard, Thomas 173, 174, 177
Hufbauer, Gary Clyde 82
Hui Liangyu 99, 100
human rights 21, 34, 48, 51, 67, 72, 88, 123, 197, 199
humanitarian aid 4-5, 115, 180
 misuse of 182-183

IAEA (International Atomic Energy Agency) 8, 67
 inspections 4, 17, 19, 34, 35, 39, 54, 136
 monitoring 37, 161, 169
ICBMs 195
IGS (Information Gathering Satellite) 124, 127, 130
Illicit Activities Initiative 68
IMF 181
India 29, 49-50
individual level analysis 90-93
defined 90
integration theory 192
intelligence-gathering 124, 127, 153
intergovernmentalism 193
international relations (IR) theory 76
Iran 4, 27, 31, 49, 66, 80(*n*59), 81, 94, 141, 201
 UN Security Resolution on 107-108
Iraq 29, 47, 66, 68, 71, 77, 81, 84, 90, 94, 98, 129, 201
Islamic terrorism 4, 31, 47, 78
isolation policy 68, 70, 71, 73, 76, 84, 86, 91, 92(*n*130), 93, 159, 164
Israel 31, 94
issue linkages 80-82
Ivanov, Igor 145(*n*44), 147
Ivashentsov, Glev A. 139, 140, 143

Japan 1, 11, 67, 74, 92(*n*130), 103, 111-131, 199
 abduction issue 21, 111, 112, 114, 116-117, 118, 119, 122, 123, 125, 131, 153
  primary importance of 128, 129-130, 144(*n*41)
 and ACSA 116
 aid to DPRK 114, 117, 121
 bilateral talks 117, 118, 122, 128
 and China 95, 96, 97, 109, 128, 168
 Chosen Soren 122-123
 colonial period 45, 50, 95, 112, 168

comprehensive security paradigm of 10
defence/military issues 11, 116-117, 124-125, 171, 197
BMD project 119, 124, 130, 199
hardliners in 130
legislation 117
PSI training 119
Foreign Exchange & Foreign Trade Law (2004) 115-116, 123, 124
and KEDO 177, 179, 182
LWR issues 111, 114
missile issue 112, 114, 117, 118
response to tests 123-127, 142, 143
normalization issue 111-112, 122
North Korean Human Rights Law (2006) 123
and nuclear deterrence 4
Pyongyang Declaration (2002) 5
response to HEU program 113
and ROK 128
and Russia 127, 128, 134, 135, 147-148
sanctions imposed on DPRK 3, 22-23, 115-116, 199
financial 122-123
shipping 126
as security concern 191
shipping restrictions law (2004) 117
in six-party talks 20, 101, 102, 113-116, 113-123
First Round 114-116
Second Round 116-117
Third Round 118-119
Fourth Round 119-121
Fifth Round, First Session 121-123
Fifth Round, Second Session 127-128
spy satellites 124, 127, 130
unilateral actions with DPRK 2
and US 10, 27, 30-31, 57-58, 68, 113(*n*9), 129, 131, 155, 161
KEDO 173
military cooperation 116, 117
in trilateral talks 113
and war on terror 129
working group proposal 122
Jervis, Robert 77, 77(*n*45)
Jiang Zemin 97-98, 160
Joint Statement (2005) 2, 4, 13, 72, 75, 82, 84, 87, 103-104, 107, 138-139, 156-157, 162, 191-192, 201
and Bush Doctrine 92
four areas of concern 156
six clauses 103
vagueness of 163-164
Joseph, Robert 6, 75, 84, 87, 93, 185-186
Juche legacy 105
Junichiro, Koizumi 147-148

Kang Sok-ju 154, 176, 185
Karasin, Grigori 134
KEB (Korea Exchange bank) 22
KEDO (Korean Peninsula Energy Development Organization) 9, 12-13, 16, 40, 74, 113, 120, 157, 161, 167-190
background /context to 172-176
chronology of 187-190
early period 176-181
contractual problems 183
cooperation in 179
costs of/funding for 176, 178, 179, 182, 186
burden sharing in 173, 177
HFO project 178, 179
and KEPCO 181
mistrust within 177-178, 180-181, 182-184, 186
spent fuel project 169, 174, 175-176, 178, 181
training in 178
demise of 122, 182-187
Keizai Dantai Rengokai 130
Kelly, James 15, 67-68, 69, 85, 101, 153, 154, 160, 185
KEPCO (Korea Electric Power Company) 181
Kerry, John 90, 195(*n*7)
Khan, Dr A.Q. 39
Kim Dae-Jung 58-59, 64, 152, 170, 173, 179, 181, 181-182
and Kim Jong-Il 171, 184
Kim Il-Sung 51, 168
Kim Jong-Il 2, 3, 8, 9, 20, 29, 30, 35, 40, 136
and China 23, 99, 101, 105, 165
contempt for 91-92
and Japan 117, 147-148

and Kim Dae-Jung 171, 184
labeled "tyrant" 74, 184
Putin on 147
Kim Jong-Il government 47
Kim Jong-u 174
Kim Kye-gwan 6, 8, 33-34, 35, 43
and Christopher Hill 36-38, 144, 145, 156, 163, 165, 197
denies HEU program 185
Kim Yong-sam 170, 177
Koizumi, Junichiro 117, 153
Korean peninsula 9, 15, 103-104, 202-204
China in 95, 100, 103, 108
Denuclearization Declaration (1992) 19, 72, 171, 185, 191
denuclearization road map 39-40, 41, **41-42**, 106, 139, 148
history of "Korea problem" 167-172
as security concern 191
US presence on 196
Korean reunification 1, 3, 50, 54, 60, 64, 135
Korean War 10, 45-46, 50-51, 168, 181
China and 95
peace treaty for 40, 63-64, 101, 103, 120, 163
Krauthammer, Charles 69
Kumho LWR facility 181, 185, 186

Lavrov, Sergei 139, 148
Lebanon 94
Lewis, John W. 9
Li Gun 24
Libya 81-82, 155
Litwak, Robert 69
Liu Jianchao 100
Losyukov, Alexander 137, 138, 147(*n*52)
Lugar, Richard 85
LWRs (light-water reactors) 12, 16-19, 20, 39, 103, 111, 121
DPRK's right to 16-19, 72-73, 139, 147, 157, 176-177, 201
and nuclear weapons 139-140
see also KEDO

Macau *see* Banco Dalta Asia
McCormack, Sean 24
Machimura, Nobutaka 121
"malign neglect" 66
Manchuria 95
Manicheanism 83, 91
Mann, James 90-91
media 89
mediation 3
Middle East 30, 31, 66, 98
missile defence (MD) 31
missile tests 27-28, 38, 48-49, 66, 94, 101, 111, 112, 141-142, 164, 171, 194
DPRK's moratorium on (1999) 111, 112, 117, 184
see also nuclear tests; *and see under specific countries*
Mizuho Bank 23
money-laundering 6, 22-25, 38, 71, 140, 164
Asia-Pacific Group on (APG) 25
see also BDA
Mount Geumgang tourist resort 3, 30, 198

Najin-Sonbong 175, 179
nationalism 191
neo-cons 6, 47, 53, 84, 183
neofunctionalism 193, 194
New Zealand 179
NGOs 180
Nihon, Keizaidantai Rengokai 130
9/11 attacks 47, 66, 67, 152-153, 158
Nishibe, Susumu 127
Nodong missile 111, 112, 123, 195
non-proliferation 30-31
see also NPT
normalization issue 111-112, 122, 156
North Korea (DPRK)
anti-terrorism stance of 153
in Cold War 46-47
economy 104-105, 158
energy crisis in 179
food shortages in 51, 180
foreign investment in 20, 23
formation of 45
human rights in 21
in international community 62-63, 104-105, 140, 180
and Korean reconciliation 1, 3, 50, 54, 56, 60, 63, 64
in nuclear club 49, 156, 166
sense of insecurity in 195-196
standard of living in 32
tactics/strategy 159-164

see also brinkmanship
unilateral statements/Memorandum (2005) 162-163
North Korean first nuclear crisis (1993-4) 12, 15, 81, 152
North Korean second nuclear crisis (2002-3) 136-137
history of 1, 2-4
North-South Denuclearization Declaration (1992) 19, 72, 171, 185
North-South Korea Summit (2000) 171, 184
Northeast Asia 96
cooperation in 204-205
theories of 192-194
security of 1, 3-4, 9, 79, 97, 131
impact of missile tests on 30
and six-party talks 5, 7, 191-205
US influence in 98
Northeast Asia Cooperation Dialogue 204
Northeast Asian trade 11
NPT (Non-proliferation Treaty) 12, 17, 19, 39, 54, 121, 139, 169
DPRK returns to 147
DPRK withdraws from 67-69, 92, 111, 136, 154
Nuclear Agreement (six-party talks, 2007) 4-9
Bush's flexibility in 5-7, 71-72
key points 4-5
unresolved issues 7-9
nuclear club, DPRK in 49, 156, 166
nuclear energy *see* LWRs
nuclear facilities 7-8, 34-35
see also Kumho; Shinpo; Yongbyon
nuclear materials/technology 67
nuclear proliferation 4, 30-31, 81, 197
nuclear tests 1, 2-4, 12, 15, 28-33, 48-49, 55, 66, 94, 105, 142-143, 164-166, 192
China's reaction to *see under* China
implications of 3-4, 81
motives for 29-30, 38
ROK's reaction to 28, 30, 142, 143
Russia's reaction to 141-143
suspicions surrounding 182-183
US reaction to 29-30, 74-76, 86, 151
nuclear weapons 8, 35-36, 112, 120-121
disablement/dismantlement of 4, 5, 19, 20, 40-41
DPRK's capability 194-195
DPRKs motivation for 195-196
and LWRs 139-140
see also nuclear testing

Oh Kwang-chul 35
"ostrich engagement" 10, 65

Pacific War 95
Pakistan 19, 29, 182
Palestinian crisis 94
Patriot Act (2001) 22, 71, 140
"peace & prosperity" policy 59-61, 62, 70, 191, 202
Pearl River Delta 105
Perminov, Anatoly 141-142
Perry, William J. 89, 171
plutonium 8, 16, 101, 111, 156, 194
Powell, Colin 158
power stations, coal-fired 174
pragmatism/realism 83, 84, 87, 93, 97, 98-99
preemptive paradigm 29, 47, 49, 77, 89, 184
prospect theory 83
PSI (Proliferation Security Initiative) 3, 10, 32, 68-69, 75, 78, 92, 143
training 119
public opinion 90
Pulikovsky, Konstantin 147
Putin, Vladimir 12, 133-134, 136, 141, 142-143, 147
PVOC (Private Voluntary Organization Consortium) 180
Pyongyang Declaration (2002) 5

Qian Qichen 101
Quinones, Kenneth 173, 174, 176

Razov, Sergei 146
realism *see* pragmatism
reconciliation, Korean national 1, 3, 50, 54, 58
regime change policy 52, 57, 84, 98, 158, 162
Rice, Condoleezza 6-7, 19, 20, 26, 80, 91-92, 106, 121, 156, 163, 165
RMA (revolution in military affairs) 77
Rogachev, Igor 145
rogue state 196

Roh Moo-hyun 10, 17, 30, 53-54, 56, 57, 101
 liberal policy of 58-62, 64
 vulnerability of 80
ROK *see* South Korea
Rumsfeld, Donald H. 5, 87, 93
Russia 1, 3, 11-12, 74, 127, 128, 133-149, 134, 135, 147-148, 179, 199-200
 and China 109, 134, 135, 146, 149
 and counterfeiting issue 140-141
 energy aid to DPRK 138, 140
 Far Eastern region 134
 foreign policy goals 134-136
 and Japan 127, 134, 135, 147-148
 and Korean Peninsula 134-135
 and Korean unification 135
 military/defence issues 135, 149
 reaction to missile/nuclear tests 141-143
 in six-party talks 20, 82, 101, 136-141, 144-146, 148-149
 and South Korea 135-136
 and UN 27, 141, 143, 146, 148
 and US 134, 135
 working group with DPRK 138
Russo-Japanese War (1905) 95

sanctions 2, 3, 9, 15, 21-25, 27, 31-34, 66, 71, 75, 92(*n*130), 197, 201, 202
 China and 32, 105, 106, 198
 DPRK'S criticism of 121
 effectiveness of 31-33
 nine commandments of 82
 positive/negative 82-83, 91
 theory 82-83
 trade 122
 see also under Japan; United States
Sasae, Kenichiro 128, 128(*n*97)
satellites
 commercial 147
 spy 124, 127, 130
Saudi Arabia 4, 31
Saunders, Paul J. 89
Scud missiles 123, 195
Shen Dingli 99(*n*10), 106-107
Shinpo nuclear facility 167, 169, 177, 201
Shinzo Abe 106
Shiozaki, Yasuhisa 125
"shock & awe" 77
Sinpo project 40
six-party talks 1, 2, 3, 4-9, 13, 15-44, 54-56, 69-70, 86, 101-104
 Fourth Round 16-20, 71-72, 102
  Chinese compromise 18-19
  DPRK position 16, 19
  joint statement of 43-44
  key points 18-19
  ROK position 17, 18
  US position 16-18, 19-20
 Fifth Round 21-27, 33-39, 75-76, 94, 165
  China's position 107-108
  DPRK's position 33-34
  financial issue 34, 35
  sanctions issue 21-23, 28, 32-33
  three point plan 21
 "1-2-3" lineup 70
 denuclearization road map 39-40, 40, 41, **41-42**, 106, 139
 Joint Statement (2005) *see* Joint Statement
 Nuclear Agreement *see* Nuclear Agreement
 power balance in 12, 82
 and security cooperation 191-205
 working group 138
 see also under individual countries
social ignorance/social contact theories 192
soft power 10, 61-62
Sokolski, Henry 73
Song Min-soon 21, 38, 43, 53
South Korea (ROK) 1, 3, 7, 54-55, 58-59, 67, 74, 182, 198-199
 airplane bombing incident (1987) 111(*n*1)
 and China 56-57, 95, 96, 98, 107, 109
 energy aid to DPRK 4-5, 16, 17, 102, 137-138, 144, 175, 201
 food/fertilizer aid to DPRK 3, 4, 28, 30, 180
 misuse of 182-183
 in international community 52-53, 62
 and Japan 128
 and KEDO 177
 military stand-off with DPRK (2002) 153
 and nuclear deterrence 4
 peace & prosperity policy 59-61, 62, 70, 191, 202

political parties 80
reaction to missile/nuclear tests 28, 30, 142, 143
reconciliation efforts with DPRK 1, 3, 50, 54, 56, 58, 60, 63, 64
and sanctions 32
in six-party talks 82, 101, 102, 104, 114, 116
submarine incident (1996) 180-181
"sunshine policy" *see* "sunshine policy/diplomacy"
Unification Ministry 17
unilateral actions with DPRK 2
and US 3, 10, 25, 30-31, 51, 57-59, 77-78, 155, 161
alternative paradigm for 57-62
divergent policies 48, 52-54, 79-80
KEDO 173
military presence 46, 53, 74, 80, 100
sovereignty 72, 102, 160, 163, 168, 193
Soviet Union 45, 46, 50, 91, 108, 168, 175, 176-177, 194
spent fuel project 169, 174, 175-176, 178, 181
spy satellites 124, 127, 130
state level analysis 83-90
Congress 87-89
defined 83
factional politics/two level games 83-87
media/think tanks 89
political culture 83
public opinion/electoral politics 90
"strategic flexibility" 59, 64, 72-73
Strategy & Management magazine 102
"sunshine policy/diplomacy" 58, 61, 66, 85, 96, 102, 170, 181-182, 184, 202
supernotes 21-25
system level analysis 76-83
alliance politics 79-80
defined 76
issue linkages 80-82
sanctions theory 82-83

Taepodong-I missile 101, 111, 195
Taepodong-II missile 1, 2, 27, 123, 141, 182
Taian Glass Factory 104
"tailored containment" 68, 69
Taiwan 4, 31, 97, 99, 107, 197
Tang Jiaxuan 106, 164-165
Thailand 181
think tanks 89
Three Gorges Hydroelectric power station 105
TMD (theatre missile defence) 97
Tokyo-Mitsubishi UFJ, Bank of 22-23
tourism 82, 105, 198
trade 82, 105, 106
sanctions 122
transfers 82
transparency 154, 205
Turk, Lynn 173, 174
Turkey 4, 31
tyranny, DPRK as 163

United Nations (UN)
Charter
Article 41 3, 123-124, 142
Chapter VII 123-124, 125-126
Security Council 66, 69
Resolution 1695 2, 27-28, 29, 74, 88, 94, 124, 141, 148, 164, 195
Resolution 1718 3, 31, 74-75, 87, 94, 100, 106, 126, 142, 143, 146, 148, 164, 199, 201
Resolution 1737 107-108
United States (Bush administration) 1, 2, 4, 5, 7, 9-10, 10-11, 12-13, 26-27, 40, 45-64, 111, 137-138, 151-166, 184-186, 196, 196-197
ABC policy 152-154, 184-185
adversarial relations 46-51
aid to DPRK 180
alternative paradigm for DPRK 57-64
appraisal of policy 65-94, 165-166
effectiveness 54-58, 143
in second term 70-74, 197
six assumptions 157-159
three levels of analysis *see* individual level; state level; system level
"benign neglect" policy 26
bilateral talks 9, 15, 16, 26, 27, 28-29, 32, 34, 76, 102, 127-128, 144
on financial issue 36, 38-39
see also Agreed Framework
and China *see under* China
Congress 87-89, 170, 172, 182-183
currency 22
and CVID *see* CVID

elections in 89(*n*104), 90, 94
foreign policy disparities 49-50, 83-87
hegemony of 77-78, 80-81, 84, 91, 99
Iran, North Korea & Syria Non-Proliferation Act 71
and Japan *see under* Japan
and KEDO *see* KEDO
and Middle East 66, 98
missile tests, reaction to 28, 74-76, 86
North Korea Non-Proliferation Act (2006) 75, 88
North Korean Human Rights Act (2004) 48, 88
and Nuclear Agreement (2007) 5-7
and nuclear tests 29-30, 74-76, 86, 151
Patriot Act (2001) 22, 71, 140
policy review (2001) 66-67
public opinion in 90
regime change policy 52, 57, 84, 158, 162
"regional" approach 154, 158
sanctions 2, 3, 9, 15, 21-25, 27, 33-34, 66, 71, 75, 113, 118, 136, 197
DPRK's criticisms of 121
effectiveness of 31-33
separate from nuclear issue 128, 144, 157
in six-party talks 16-18, 19-20, 114, 116, 154-157
and South Korea *see under* South Korea
trilateral talks (2003) 154-155
and UN 74-75, 125-126
see also Clinton administration
uranium *see* HEU
Uri party 80
US-DPRK Agreed Framework (1994) *see* Agreed Framework

Van Ness, Peter 77
verification 19, 20
Vershbow, Alexander 91-92

Wall Street Journal 89
Waltz's three levels of analysis 76
Wang Zhongwen 102
war on terror 1, 3, 4, 47, 52, 67, 77, 78, 90, 129
weak states, tactics of 165-166
West Sea stand-off (2002) 153
WMD (weapons of mass destruction) 22, 68-69, 75, 119, 124, 134, 184, 203
Wolfowitz, Paul 183-184
Wolfstahl, Jon B. 195(*n*7)
"word for word, action for action" 156, 162, 163
Wu Dawei 33, 43, 106, 146

Yan Xuetong 99(*n*10)
Yeltsin, Boris 137(*n*9)
Yongbyon nuclear facility 4, 5, 6, 16, 37, 68, 144, 161, 169
US at 174, 176, 178

Zelikow, Philip D. 26-27
Zhu Rongji 97